# ANIMAL SQUAD

Sid Jenkins and Paul Berriff

BBC PUBLICATIONS

Published by BBC Publications
A division of BBC Enterprises Ltd
35 Marylebone High Street
London W1M 4AA

ISBN 0 563 20521 0

First published 1986


Typeset by Wilmaset, Birkenhead, Wirral
Printed and bound in Great Britain by
Richard Clay PLC, Bungay, Suffolk

# CONTENTS

To Tom Taylor for his encouragement and Sue for her support

To my wife Micky and my children Lisa and John-Paul for their tolerance and understanding during the writing of this book

# PREFACE

On 10th August 1983 Paul Berriff walked into the office of RSPCA Chief Inspector Sid Jenkins at the Group Communications Centre in Leeds. Paul is a documentary film maker, and was there to cover a story for the BBC television news programme *Look North*. With him was news reporter Judith Stamper and their assignment, that summer's day, was to cover a disturbing story about the discovery of the charred remains of a cat, thought to have been the victim of a ritual sacrifice. Other journalists that Chief Inspector Jenkins had contacted had shied away from the story – they were worried that publishing pictures of the cat would be of too disturbing and horrific a nature for the public to contemplate. However, Paul Berriff and Judith Stamper quickly became certain that this was a story that needed to be brought to people's attention, and their subsequent news item appeared on that evening's programme.

Like most of us, until that moment Paul – if he'd thought about it much at all – had assumed that the RSPCA is an organisation mainly concerned with running homes for stray dogs and cats. However, his discussions with Sid Jenkins soon impressed on him just how diverse and extensive the RSPCA's investigative work is – and how professionally its inspectors carry it out. It was a side of their work he had no idea of at all, and Paul was fascinated to learn from Sid that each region of the country is divided into teams of RSPCA inspectors who operate on similar lines to the police, with all the methodical and painstaking investigation that that implies – even, as Paul discovered from Sid's stories, stake-outs, road blocks and undercover work. When Sid informed him that the number of calls coming in to the RSPCA reporting cases of cruelty to animals – some of quite horrifying dimensions – is increasing at a truly appalling rate each year, Paul's experience of making television documentaries immediately told him that here was a subject that deserved and lent itself to a television film.

The idea gradually took seed, and over the next eighteen months Paul spent more and more time with Sid researching all the various aspects of the RSPCA Inspectorate's work. He specialises in making 'photojournalistic' documentaries, and it quickly became apparent to him that Sid's enormous patience and strength of personality, as well as the professionalism he brings to his work, would make him an ideal subject for a film of this type. Once their discussions were advanced enough for Paul to have compiled a synopsis for a programme, they approached Charles Marshall, the Chief Officer of the RSPCA's Inspectorate, Peter Jarvis, its legal officer, and Mike Smithson, Director of Publicity, to explain what they wanted to do. Jointly they gave the go-ahead for the project. Paul Berriff next approached the BBC, who liked the idea and commissioned him to make a fifty-minute documentary special for BBC 1.

In the late summer of 1985 the film team joined Sid Jenkins and his colleagues to start work on the project. It was to be an unprecedented period in the history of Group 2, Region 8 of the RSPCA Inspectorate. Indeed, throughout the country during these months the RSPCA was inundated with a record number of complaints to investigate and prosecute where possible. It quickly became obvious to Paul that he had far more stories here than could be fitted into a single programme. When Roger Mills, Head of

Documentaries for BBC 1, saw the first 'film rushes' he agreed. Paul now had the go-ahead to make a series of six thirty-minute films.

This book was written to complement the television series. In a film lasting only thirty minutes a lot of background detail has to be omitted. The book can go further into the details, and can follow up certain of the cases after the television camera has ceased to cover them.

Sid Jenkins and Paul Berriff would not have been able to make the films without the tremendous help and co-operation of very many people, including Charles Marshall, Chief Officer, RSPCA Inspectorate, Regional Superintendant Len Flint, and Inspectors Dave Millard, John Oxley, Kevin Manning, Derek Woodfield and Steve Wilkinson of Group 2, Region 8, as well as communication controllers Elaine Mitchell and Trish Brown, Alec Hollingsworth, Secretary of the Leeds and District Board of the RSPCA, the manager and staff of the Leeds RSPCA, and John Saxton, Veterinary Surgeon. Without the foresight of Roger Mills and the dedication and professionalism of the film team – Ray Parker and Jon Pinkney – who worked in some appalling and very sensitive situations, it would not have been possible to capture the front-line dramas of the RSPCA's 'animal squad'.

# INTRODUCTION

On the evening of 16th June 1824 a number of distinguished people, including the MPs William Wilberforce and Richard Martin, assembled at Old Slaughter's Coffee House at the corner of St Martin's Lane and Cranbourn Street in London. The meeting had been called at the instigation of the Reverend Arthur Broome, vicar of the parish of St Mary's, Bromley, in the east end of London. As a result of their deliberations that evening, a society was born, its sole purpose to prevent cruelty and promote kindness to animals. That society later became known as the Royal Society for the Prevention of Cruelty to Animals, the first animal welfare organisation in the world.

William Wilberforce, a Yorkshireman, had already won renown as a humanitarian as the result of his hard-fought campaign to end the slave trade in the British colonies (the first attempt to abolish slavery entirely had gone before Parliament only the year before). He had now turned his attention to the suffering of animals, in an era when bull baiting and cock fighting were common pastimes, and it was not unusual to see a horse flogged to death in the streets.

Along with Lord Erskine, Wilberforce and Martin took action to introduce into Parliament a previously unheard-of measure. This was 'A Bill to prevent the cruel and improper treatment of cattle' which proposed 'that if any person or persons having the charge, care or custody of any horses, mare, gelding, mule and ass, cow, ox, heifer, steer, sheep, or other cattle, the property of any other person or persons, shall wantonly beat, abuse or ill-treat any such animal, such individuals shall be brought before a Justice of the Peace or other magistrate.' The penalty was fixed at a fine not exceeding five pounds nor less than ten shillings, or imprisonment for not more than two months. The Bill duly became law, and was thus the first piece of legislation aimed at the protection of animals passed by any parliament in the world.

Support for the animals' cause had come not only from politicians but also from artists and writers of the time who were concerned about the brutalising effects that acts of cruelty to animals had on people. The plans drawn up at that first meeting of the RSPCA included a strategic campaign that is still broadly adhered to today. Printed matter on the care and welfare of animals was to be produced and freely distributed, frequent appeals for support would be made to the public through the press, and inspectors would be employed in the markets and streets to check on the condition of animals and, in the case of flagrant acts of cruelty, to prosecute the offenders. Even before that first meeting was called, the Rev Arthur Broome had employed a man, a Mr Wheeler, and paid him out of his own pocket, to keep a watchful eye on acts of cruelty. After the Society had been formed a second inspector was appointed, and between them in their first six months they brought 63 offenders before the courts.

This, then, was the humble beginning of the RSPCA Inspectorate, which today consists of 240 uniformed inspectors, stationed throughout England and Wales. It is organised administratively and operationally from the RSPCA's national headquarters in Horsham, Sussex, and is responsible for all the uniformed inspectors and also for the headquarters' superintendents employed on prosecution work. The Inspectorate is divided into eight regions, each commanded by a Regional Superintendent, and these are further sub-divided into Groups run by a Chief Inspector.

In 1985 the RSPCA Inspectorate countrywide investigated 64,678 complaints of animal cruelty from the public. 2112 convictions were obtained in the courts, and verbal cautions for acts of cruelty, not considered enough to warrant legal proceedings, totalled 4228. The number of telephone calls dealt with by the Group Communication Centres nationwide was 790,346. Calls dealt with by inspectors out of hours numbered 217,778.

The prime function of the Inspectorate is to act as a law enforcement agency. The inspectors are the 'animal policemen' of this country and know more about animal legislation than most solicitors. The law requires in every case of cruelty that there should be proof of substantial unnecessary suffering. When legal action is being considered the inspector concerned will first supply detailed evidence of the case to headquarters where the reports will be studied by one of the RSPCA's specially trained legal superintendents who will decide whether or not to prosecute. If the legal superintendents sanction a prosecution then the inspector who originated the case is instructed to 'lay information' with the local Magistrates' Clerk, who will then issue a summons. When the case comes to court, it will normally be the inspector's evidence on which the prosecution case will rest. The inspector will also have briefed and instructed his own local solicitor to act on the RSPCA's behalf.

The principal act for the protection of animals is known as the Protection of Animals Act 1911. This prohibits all kinds of ill-treatment to animals, and anyone, whether the owner or not, who permits an animal to suffer, is held to have acted unlawfully and is subject to punishment.

Permitting cruelty means failing to exercise reasonable

care and supervision. It is generally a question of fact whether or not an owner has exercised proper care, and the negligence of the owner must be so gross as to receive the condemnation of the criminal law. A cruel act has got to be wilful: allegations that an animal has been caused unnecessary suffering have to be shown capable of proof before the RSPCA will proceed with a prosecution. Acts of omission are punishable as well as acts of commission: this means that a person may be convicted for neglecting to provide adequate care for an animal so as to cause it suffering.

When the 1911 Act was passed it was welcomed as the Animals Charter, and is still so regarded today, as it provides the framework that allows the Society's inspectors to grapple with cases of cruelty. The part of the 1911 Act most used by them is Section 1; which establishes the various offences of cruelty, provides the penalties and gives some qualified exceptions to the general protection given to captive and domestic animals, as follows:

1. If any person
   (*a*) shall cruelly beat, kick, ill-treat, over-ride, over-drive, over-load, torture, infuriate, or terrify any animal or shall cause or procure, or, being the owner, permit any animal to be so used, or shall, by wantonly or unreasonably doing or omitting to do any act, or causing or procuring the commission or omission of any act, permit any unnecessary suffering to be so caused to any animal; or
   (*b*) shall convey or carry, or cause or procure, or, being the owner, permit to be conveyed or carried, any animal in such manner or position as to cause that animal any unnecessary suffering; or

(*c*) shall cause, procure, or assist at the fighting or baiting of any animal; or shall keep, use, manage, or act or assist in the management of, any premises or place for the purpose, or partly for the purpose, of fighting or baiting any animal, or shall permit any premises or place to be so kept, managed, or used, or shall receive, or cause or procure any person to receive, money for the admission of any person to such premises or place; or

(*d*) shall wilfully, without any reasonable cause or excuse, administer, or cause to procure or being the owner permit, such administration of, any poisonous or injurious drug or substance to any animal, or shall wilfully, without any reasonable cause or excuse, cause any such substance to be taken by any animal; or

(*e*) shall subject, or cause to procure, or being the owner permit to be subjected, any animal to any operation which is performed without due care and humanity;

such persons shall be guilty of an offence of cruelty within the meaning of this Act, and shall be liable upon conviction to a fine and/or imprisonment.

The stories in this book relate to a number of incidents that took place over a period of two or three months in the autumn and winter of 1985/6 in Group 2 of Region 8 of the RSPCA's Inspectorate, an area which covers most of West Yorkshire and part of North Yorkshire. The team making up this Group consists of Chief Inspector Sid Jenkins and five other inspectors. Sid Jenkins is stationed in Leeds and is responsible for investigating complaints of cruelty in this area, as well as supervising the day-to-day running of the

Group and giving coverage to his other colleagues. Inspector Dave Millard, stationed at Bradford, deputises for Sid when he is on leave or away on duty. Inspector John Oxley is based at Skipton in the Yorkshire Dales. This area is mostly rural, among some of the most beautiful scenery in the country, and his work takes in farms and cattle markets, but he will also provide coverage for his colleagues in the cities when necessary. Inspector Kevin Manning, stationed at Huddersfield, has a mixture of town and country work. His patch includes the town of Holmfirth, the location for the television series *The Last of the Summer Wine*. Inspector Derek Woodfield operates closely alongside Kevin Manning at nearby Halifax. Finally, Inspector Stephen Wilkinson is stationed at Harrogate, covering the town and surrounding North Yorkshire countryside.

The Group Communication Centre is based in a converted mansion house in Chapel Allerton, Leeds, and is manned by Elaine Mitchell and 'Trish' Brown who work a one-week on and a one-week off system between them. They receive all the telephone calls for the Group and pass them on by radio to the appropriate inspector. This relieves the inspectors from the need to sit close to the end of a telephone, and allows them much greater mobility to respond quickly to urgent calls. The base in Leeds now receives up to 100 calls per day. Last year they answered over 19,000 calls, which led to 2500 serious investigations, as a result of which 100 people were successfully prosecuted and convicted in the courts.

We like to pretend that we are a civilised nation, that the horrors the RSPCA was founded to oppose are buried back in our Victorian past, along with the gin parlours and boy chimney-sweeps. Regrettably, it's not true. In the 1980s nothing much has changed. There is still horrifying cruelty towards animals going on under our very noses. Not only in the back streets of our large cities, but in the well-heeled, prosperous housing estates on the outskirts and in the pretty villages deep in the countryside.

The accounts given here illustrate the kind of thorough investigative work which is part of the day-to-day life of Chief Inspector Sid Jenkins and his team. He operates a unique system which allows all five of his inspectors to move freely from area to area, coming together for the 'big jobs', or when back-up is required. They are in constant radion communication – even at home. Sid Jenkins's office, two rooms in a Victorian mansion in Leeds, is the main control centre. When information is received about a 'special case', he will call an immediate conference of all his five colleagues to discuss how they should set up observation and investigate the report.

Sid lives in north Leeds with his wife Sue, a former veterinary assistant. His work makes it necessary for him to be an unofficial social worker, for he deals with as many people as he does animals. 'My Inspectorate have to be tactful, tough and physically fit,' he says. 'We very rarely have problems with animals – it's people who cause the problems.' Sid is convinced that prevention is better than cure. Education rather than prosecution is what he strives for. With his pet cat Orlando, a beautiful ginger cat that he found abandoned in a ditch a few years ago, he visits schools, Women's Institutes and groups, spreading the message of the RSPCA. Not with a big stick and a threat of damnation, but whenever possible with a smile, a joke and his very own magic act – methods very much in keeping with the original intentions of the RSPCA when it was formed 162 years ago.

At the end of it all, it is Sid Jenkins and his team of

inspectors, and the others like them up and down the country, who stand between the animals and the staggering cruelty still practised by thousands of unthinking and heartless men and women in Britain today.

# EGG FACTORY

The duty roster showed that Chief Inspector Sid Jenkins was supposed to be having a long-overdue day off-duty. But the appearance of a man at Leeds Magistrates Court on a charge of causing unnecessary suffering to a number of dogs had put paid to any hopes Sid might have had of enjoying it. It was to be the start of a day that would be etched on his memory forever – one result of which would be that his eating habits would never be the same again.

Having heard Sid's evidence, the magistrate fined and disqualified the defendant from keeping dogs for a considerable time, and Sid returned to the Group Communication Centre at Chapel Allerton. It had been an appalling summer but now the weather showed signs of turning warm at last, and Sid was hoping to take the rest of the afternoon off to take his wife, Susan, out into the country. Sid was about to see the country all right, but not the way he had expected.

'How did it go in court?' asked Elaine as he entered the office. One of the Group's communication co-ordinators, she is responsible for receiving all the incoming telephone calls from the public. 'We got a good conviction,' said Sid smiling. It had been a busy year for the Group and the number of court convictions had broken all records. This was another statistic to add to the growing list. Sid is always pleased when he obtains a successful prosecution. He feels strongly that the newspaper publicity given to every case of cruelty or neglect proved in court acts as a deterrent to the public, and prevents further possible deeds of cruelty and neglect.

Just then the telephone rang. It was a call for Sid. 'Whereabouts? You think there's suffering? Right, I'll take a look.' Sid put the phone down, then picked it up again to ring Susan and tell her that he was on his way home. Not to take her out, though, but only to change from his uniform into 'civvies' to investigate this latest 'job'.

The call had said that hens were being neglected at a battery unit on a farm near a tiny village nestling under the shadow of Almscliffe Crag, one of Yorkshire's many beauty spots in the picturesque Wharfe valley, only a few minutes from Sid's home. Quickly there, he was careful to park his van down a lane about half-a-mile from the farm's entrance and walk the distance there – he couldn't afford to blow his cover before he'd even started the investigation. For this sort of operation, the vans the inspectors use have detachable RSPCA signs – usually on the side. Before he had left his office on this enquiry, Sid had taken them off the van.

The farm lay on the busy Leeds to Skipton road, the gateway to the Dales, and the farmer had erected a large sign advertising 'eggs for sale'. Sid gazed up at the sign and then strolled down the tree-lined track towards the farm. The approach was littered with the remains of wrecked cars. There was also a boat, a cooker and farm machinery that had all been left to rust in the elements. It was little short of a scrap-yard. As Sid approached the farmhouse a couple of dogs, confined to the compound, started to bark.

Sid's story was that he was there 'to buy some eggs'. He knocked on the farm door, but there was no reply. The farmyard, next to the back door, was strewn with debris. There were empty egg boxes, clothing, old shoes, cardboard boxes and plastic containers all over the place. Sid leant against the five-bar gate that led into the yard, and gazed at the desolation in front of him. All around were dilapidated buildings.

'Hello,' shouted Sid, 'hello. Anyone about?' There was no response. At the bottom end of the farmyard he could see a number of hen houses. Their roofs had collapsed. He climbed over the gate and continued 'looking round'.

A horror-movie film had nothing on this. Sid walked along slowly absorbing the dereliction all around him. There were buildings with no roofs, buildings with no side walls, buildings that had collapsed like packs of cards. Everywhere there were smashed and broken farm implements. Sid could hardly believe what he was seeing. To his left he came across a pen constructed out of wood and wire. Glancing inside he saw the carcases of ten hens in a state of advanced decomposition. They had obviously lived and died in the cage. He stared down at the hens and tried to speculate how long it had taken the poor birds to die.

Next, he moved across to look more closely at one of the collapsed units. Climbing up onto the broken wooden sections for a better view, he could see that the roof timbers had fallen down onto the cages of the battery hen units. Large numbers of dead birds, again in a state of decomposition, were still inside. Sid was totally sickened, and very angry, at this evidence of appalling neglect by a farmer who appeared only concerned to make money out of the birds supposed to be in his care.

Chief Inspector Jenkins is not easily shocked. He'd seen it all before, or so he thought. But this was something else altogether. He back-tracked up the farmyard. On his way in he'd noticed a door leading into one of the battery units. 'Hello,' he shouted, approaching it now. There was still no response. The door appeared padlocked. He could hear the sounds of live hens inside the building. He was beginning to get quite agitated by now, and was determined to look inside.

All around the farm and the buildings were weeds and overgrown vegetation – the only clear bit of space was the concrete path on which he was standing. Now the only thing that Sid could do to reach the back of the buildings was to plough his way through the undergrowth. The nettles growing at the rear of the units were taller than Sid, but that wouldn't stand in his way. Using his arms as scythes, he pushed his way through, ignoring the stings, his mind on one thing only: the condition of those birds in the sheds. At last he reached the back of the main unit, but his problems were still not over. He began to sink. All at once he realised that this was a place where an investigator should tread carefully – the smell was enough to tell him that. The farmer must have been shovelling the slurry for years straight out of the back of the units. It lay in a stinking, stagnant moat around the base of the sheds. Dotted around were sacks containing the remains of dead hens piled high on top of one another. Where the back door had once been there was now a large piece of hardboard, leaving a gap of two feet through which the inside of the shed could be viewed. But before he could get there Sid was going to have to climb the mound of slurry, heaped up six feet high, a fermenting Everest straight in front of him.

Suddenly he started to sink even further into the sticky, nauseous mess. Somehow or other he managed to grab hold of a plank of wood lying nearby, and used that to heave himself up the mound. After a bit of a struggle he was standing on top and peered inside. What greeted him was a scene of unimaginable devastation and chaos.

An intensive poultry-rearing unit is a single-storey building, approximately 100 yards in length. It houses thousands of hens. They are kept in wire cages which run the length of the building and are stacked up to three tiers high. Water and food is provided by way of plastic pipes and a conveyor belt system, which provides food for the birds at fixed times of the day and is attached to the front of the cages. Most of the eggs that are bought in the shops today are produced by this system. The hens sit in small wire cages and lay eggs which roll to the front of the cages and are then collected by the farmer.

But that wasn't what Sid was seeing. 'Christ Almighty.' His heartbeat started to race, and he was breathing heavily. He lifted the piece of hardboard from the doorway and stepped inside. An overpowering smell of ammonia wafted down out of the shed and mingled with the fresh country air of the summer evening. Sid's brain was struggling to grasp the extent of the carnage he could see all around him. There was no automation at all. The birds were absolutely bare. He'd seen enough. He turned round and made his way back to the fresh air of the doorway. This beat anything he'd seen before in all his years as an RSPCA Inspector. The number of dead birds was unbelievable. He couldn't believe that anybody could neglect fowl like this. There was absolutely no excuse for it.

Sid breathed deeply, grateful for the clean air in his lungs and the chance to clear his thoughts. This was a major incident. He would need to get hold of the police, the Ministry of Agriculture and a vet. They must all come out and see it for themselves. But first he would make one last attempt to raise the farmer responsible for this heartless suffering. Jumping down through the pile of slurry, he hurried back through the nettles and up the track to the farmhouse. He banged long and loudly on the door, seizing the chance to vent his fury. Still there was no reply. But there was no time to waste. No matter that he stank of slurry, Sid leapt into his van and drove the half-mile to the nearest police station.

PC Leighton Morris was the community constable on duty. A middle-aged man who looked as though he spent most of his time out in the countryside, he had a jovial manner, and this helped to cool Sid's anger as he related the story of his findings at the chicken farm, describing the conditions in detail in a now calmer way that belied his still agitated feelings. He told PC Morris that he would be returning to the farm later that evening with a vet, and would also require police assistance to gain entry to the units and ensure that there would be no breach of the peace. The policeman appeared unsurprised by Sid's account of his afternoon's work. He agreed to give Sid all the help he could, and said he would ask his Sergeant to be present on the return visit to the farm as well. The time was now 5 pm.

Sid returned home to change from his 'civvies' back into his uniform and to organise his team for the return visit. He arranged for two colleagues, John Oxley, who was stationed at Skipton, and Kevin Manning from Huddersfield to accompany him, as well as a veterinary surgeon, John Saxton. At 6.30 pm a police car, a police van and three RSPCA vans drove down the lane and parked among the wrecked hulks of abandoned cars at the front of the farm.

Police Sergeant Fox was the first into action. He banged on the farmhouse door and received no reply. The men didn't hang about. Sid quickly led them through the farmyard to the hen units, showing them the carcases of the hens in the outside pen before they made their way to the main shed. On his earlier visit Sid had assumed that the padlock on the front door of the hen house meant that it was locked. He should have known better. Like so many things on the farm, it did not work and the door opened easily. Sid could have saved himself all his earlier time and effort, and especially his messy entrance through the slurry.

The smell that greeted them on entering the unit was so nauseating that the two police officers immediately left and came back wearing face masks. The whole party was visibly shocked at the sight before them. Sid's colleagues, both of whom were responsible for large farming areas, could not believe their eyes. John Saxton, the vet, had been called to many situations involving the suffering of animals over the years, but none like this. He was speechless.

Slowly, by the light of the 40-watt bulbs that was all there was to illuminate the whole complex, they began to make out the horrifying details of the scene. They were standing in a wooden building constructed like the old army huts known as 'Spiders'. This could not have been a more apt description of the unit they were in: the whole place was covered in large cobwebs. The door at the rear of the unit was missing and there were large gaps in the roof so that the live birds were exposed to wind and rain. Three rows of back-to-back cages, three tiers high, ran the length of the building, one at each side and one down the centre. There were no live hens at all in the cages of the first row, but close inspection revealed the full horror of what was there. Each one contained at least two or three dead birds in various stages of decomposition.

'Every cage – look.' Sid Jenkins was stunned. Even on the floor, dead birds were piled three and four deep. Bones crunched underfoot as the men moved gingerly around.

'We're walking on dead birds,' exclaimed PC Morris in disbelief.

'They've just been left to die,' said Sid. 'Some more recently than others.' He pulled out a dead bird from where it had been trapped by its neck under the wiring of a cage.

Inspector John Oxley had found a sack full of carcases. 'A bagful here,' he shouted.

'Right,' said Sid, grimly determined, 'we're going to take each body out and count them. We have literally hundreds, thousands of dead birds. That bird was trying to get out underneath the cage. It died with its neck out, poor thing. Look how some have died trying to get out. I must admit I'm beginning to feel a bit angry.' That was some admission from the usually even-tempered Chief Inspector.

There were birds everywhere. The men had reached the end of the first aisle. They turned to come back up the other side. 'God Almighty,' roared Sid, 'there are more down here. One, two, three . . . six in that cage.' Just then, there was a flapping of wings and a hen that had escaped from its cage fell suddenly to the floor. 'Try and get hold of it if you can,' said Sid to one of the police officers. The bird made no attempt to move as the policeman made a grab for it. 'Well, it didn't run away,' said Sid as he handed the bird over to John Saxton.

'It feels a bit light,' said the vet as he examined the bird closely.

'There's nowt in its crop,' explained John Oxley, taking the bird into his care.

The second row of cages mirrored all that had gone before. Some of the hens had their heads and necks caught and twisted in the grotesque positions they had been trapped in by the wire as they thrust themselves out of the cages to look for food.

The third row contained the live birds. They were living in the most shameful conditions imaginable. The extractor fans over the cages were not working and the strong smell of ammonia had become unbearable. The conveyor belt that carried away the droppings from the birds was not working either, and it was obvious that it had not done so for some time. Every single cage in the row was in a terrible state of repair. Some of the cages had paper sacks stuffed under the bottom to stop the hens from dropping through the rotten wire. Some hens had their heads caught and trapped in the bottom of the cages and had to be released.

Each cage, which measured 29 inches by 18 inches, had been divided into two sections by a metal plate. These were mainly all buckled and distorted so that the space available to the hens was considerably reduced. The metal parts of the cages had rotted and broken away, leaving jagged edges which exposed the birds to injury as they crawled over each other in an attempt to reach the water nipples.

'I mean, this really isn't on, is it?' remarked Sid to John Saxton, shaking his head in disgust.

'No, he's been sub-dividing his cages,' said the vet.

Sid went on, 'They are entitled to at least four inches of trough space, but these are big birds, what's left of them, and I've noticed six in some cages – I've even seen seven. Look, in this one there are four on either side of the divider. They're fighting each other to get to the bars. He's not been inspecting them properly.'

The vet nodded in agreement. Guidelines for keeping domestic fowl are laid down by the Ministry of Agriculture in their *Codes of Recommendation for the Welfare of Livestock*, published as a booklet. It was obvious to both men that whoever owned this unit was ignoring even the minimum requirements that this document lays down. Among these are stressed the need to make regular checks, and remove dead birds, and to avoid overcrowding of the cages.

Sid moved on to the next cage. 'Oh look at this. Now this is definitely failure to inspect. That's recent – that's very recent – the poor thing was trapped by a piece of metal. It probably fell through and the gap wasn't big enough for the bird to get out.' Sid lifted the limp body of the hen through the tiny space at the bottom of the cage.

'And that's criminal in a cage of live birds,' he said, pointing to a jagged piece of metal that looked like a bread knife.

Above the din from the live hens, there came a shout from Inspector Kevin Manning. 'Look at this – a dead bird on the floor of the cage, and the others have laid an egg on top of it.' Sid Jenkins came to look. 'It's been there so long it's formed part of the cage. It has been trampled down by its mates.'

After an hour and a half of checking and recording, and trying to relieve the suffering of the birds that were still alive, the team walked out of the battery unit. It was now dark, but they were thankful to be out in the fresh air. At ten minutes past eight a car drove down the track to the farm house. It was the farmer. There was a woman with him who turned out to be his mother. Considering the number of vehicles and uniformed people that greeted the couple, it was remarkable just how little reaction was shown by either of them. They appeared neither surprised nor upset at the reception committee. Sid Jenkins was the first to speak.

'When did you last see your hens in the battery units?'

'Which one do you mean?'

'All of them; have you seen them lately?'

'Oh yes,' said the farmer, obviously taken aback by the question.

'When did you last inspect your units then?'

'We inspect them every day.'

'And everything is OK, is it?' asked Sid, pen poised, ready to note down the answer.

'Well, the birds are quite fine, like.'

'They are, are they?' Sid Jenkins looked straight into the eyes of the farmer.

'Well, yes, they lay.'

It seemed that all the man cared about was the fact that the hens laid eggs. Sid took a deep breath. It was important that he carried out his questioning in a professional manner and remained objective. He went on, 'You follow the Intensive Regulations, do you, in accordance with the law?'

'Well, we give them water,' came the reply.

'Who's we?' asked Sid. The farmer did not answer but went on, 'We feed them every day.' Sid Jenkins persisted with his question, 'Who's we?'

'I do,' replied the farmer.

'Do you own them?'

'Yes.'

Sid then asked the farmer for his full name, address and occupation. He was hesitant in giving his occupation, but when pressed replied, 'Well, we farm, yeah.'

'Is it just you who owns the farm?' asked Sid.

'No, me and my mother own it.'

'When did you last inspect your hens?'

'We inspect them every day,' came the reply.

'When did you last inspect them?' insisted Sid, raising his voice.

'Today.'

'What time today?'

'It would be about five o'clock.'

'And did you notice anything unusual?'

'In what way?'

Sid Jenkins noticed that the replies to the questions were becoming more hesitant and more suspicious. His past experience of interrogation told him that when a question is answered with a question, it is a sure sign that the suspect is feeling insecure. The interview went on.

'Did you notice anything unusual in the cages you inspected?'

'Umm . . . well, I don't know what you're getting at.'

'Did you notice anything wrong with your hens?'

The farmer's reply should have staggered his listeners, but it didn't. After what they had seen on the farm during the past two hours, nothing could surprise them. 'Well, the hens are quite fine.' The man appeared quite unconcerned.

It was too much for Inspector John Oxley. He had been listening patiently to the questioning, but could hold back his frustration no longer. 'Didn't you see the cages the hens were in? There were lots of jagged edges and bits of metal sticking out.'

Now the woman spoke for the first time. 'I examined them yesterday and there were about five cage fronts sticking out.'

Sid had heard enough. He turned to them both in turn. 'Now, I've looked at your units and there are hundreds of carcases of hens. That means you've failed to inspect them properly. Is that correct?'

'No, I wouldn't agree with that,' replied the farmer.

'Are you telling me there are no dead birds?'

'No, I'm not saying that.'

This was the first real admission from the farmer. Sid was busily writing down all his answers in his notebook. 'You're not saying there are no dead birds?' he asked.

'We did have a problem at one time with some dead birds.'

'In amongst your live birds, are there any dead birds?'

'Not to my knowledge.'

It was at this point that the woman made another of her rare interjections. 'There might be,' she said. Her son looked astonished at his mother's contradiction.

Sid repeated her words. 'There might be?' he asked.

The woman caught her son's look and hurriedly went on, 'Well, I didn't see any yesterday.'

'But,' Sid queried, 'you said you went there at five o'clock today?'

'No, I didn't go,' she replied.

'I did,' offered her son.

'So you didn't see any dead birds at five o'clock today?'

'I didn't see any birds in the cages today.'

'So if there were dead birds there, then you didn't make a thorough inspection. How long were you in the unit?'

This was an important question. Sid Jenkins was trying to prove that under the relevant part of the *Code of Recommendation*, the couple had 'failed to make a thorough inspection' and also 'to remove dead birds promptly'.

The farmer replied, 'Well, I just did a quick feed on them.' Sid pressed on with his questions, attempting to establish just how thoroughly the farmer had looked at his hens. It was clear that he had failed to notice any dead birds in the cages that afternoon, and when Sid asked him what time of day he normally went around the cages, the farmer's nonchalant reply astonished him: 'Whenever I can fit it in.'

Sid felt himself growing angry once again, and his voice began to rise: 'So when will you fit it in?'

'Well, I'm home now.'

'So – will you go now?'

'Well, I'll have something to eat first.'

His listeners were lost for words. There were hundreds of dead birds lying in the units, and all the farmer could think about was having his evening meal.

Sid Jenkins had been pushed far enough. Deliberately and forcefully, weighing each syllable, he asked: 'When did you last make a thorough inspection? I *mean* a thorough one?'

'Well, we mucked out yesterday, didn't we?' The farmer turned to his mother for affirmation.

'Yes,' she replied.

Sid closed his notebook. 'Right, I'd like you both to come round the unit with me *now.*'

For the second time that evening PC Leighton Morris pulled back the sliding door into the number one battery unit. The nauseous stench, by now all too familiar, again hung sickeningly in their nostrils. In the evening light the shed was gloomier and spookier than ever, and PC Morris asked the farmer where the light switch was. 'The lights are on,' he replied. The beam from a bicycle torch could have generated a brighter glow, but three 40-watt bulbs were the only source of lighting in the whole building, and were all located on the right-hand side of the unit. Not a single bulb in the row of lights down the centre passage was working.

Good lighting is necessary in a battery unit, not only to allow proper inspection of the cages, but also to encourage the birds to eat and drink. If hens are allowed to remain in darkness they will believe it is a period of rest. The relevant passage of the *Code of Recommendations* states that 'Provision should be made for a period of darkness in every twenty-four hour cycle. Enough lighting should be available to enable *all* birds to be seen clearly when they are inspected.'

Sid Jenkins took the farmer and his mother down the unlit central aisle and by the light of his torch pointed out the large numbers of dead birds lying in the cages on the floor. 'What about these?' he asked.

'Well, these are birds which have died in the past.'

Sid told the farmer shortly that he could work that out for himself, and asked if he knew the regulations and codes relating to dead birds, which states that 'dead birds must be removed from the cages immediately'. The farmer did not reply.

'You would agree that this is an intensive unit, a battery unit?' asked Sid.

'Yes, that's right.'

'Well, then, why are there still dead birds here?'

The farmer then made another of his reluctant admissions. 'They should have been taken out.' Sid Jenkins felt that at last his questioning was getting somewhere.

Just then John Oxley's foot hit a paper sack. It burst open, spewing its grizzly contents all around. Once again the farmer admitted that the remains of the birds should have been taken out of the building. Sid now asked whether the farmer had had a veterinary surgeon to his hens. The farmer hedged a bit, obviously uneasy, and then admitted that he did not have a regular vet.

'When *did* you last have a veterinary surgeon on your premises, then?' asked Sid. He would not have thought it possible that he could hear anything further that evening to astonish him, but the farmer's next answer rocked him back on his heels. Standing there amongst all the carcases lying in the cages, on the floors, and piled into bags and boxes, the farmer said 'Well, we've had no problems with these birds to need to call a vet.'

'You've had no problems?' Sid pointed in disbelief to the hundreds of dead birds.

'No.'

Sid turned to the woman. 'And what have *you* got to say about these dead birds?'

'Well, I thought . . . I've nothing to say.' She had caught her son's eye once again.

'But you agree all these birds are dead?' asked Sid.

The woman just nodded her head in resigned agreement.

Flashing their torches down the rows of cages on the other side of the inspection aisle, the men started to inspect the live birds. Here the cages were three tiers high: only the top cages were in use. Each row of cages contained approximately 200 hens. As they moved along the row John Saxton, the vet, muttered out loud that he had never before, in all his experience, seen anything quite like this. The owners of the farm were shown the cage where, on their first visit, Sid had removed a dead hen from among a cageful of live ones. Three cages beyond that he pointed out to them the decomposing body of the bird that had been left for the live hens to lay their eggs on top of. Hadn't they spotted it before?

'I've obviously missed that one,' the farmer replied. A few minutes later Sid saw the woman go over to the cage, slip her hand inside, remove the egg from the festering carcase, and put it in her handbag.

Meanwhile Sid was having great difficulty in getting the farmer to understand what was meant by keeping the cages 'in a good state of repair'. He just kept repeating, dully and feebly, that the cages were old ones, so what could you expect. 'Oh come on now,' said Sid airily. 'I am a patient man, but don't tell me that you have kept up with the repairs when we have seen cages with jagged edges and tied up with bits of string.'

At that moment they surprised a couple of live hens scrabbling in the darkness on the floor of the shed at their feet. The woman remarked that all the hens had been in their cages when she had 'inspected' them earlier that evening, so they could only just have escaped. 'If the cages were in a proper state of repair,' John Oxley pointed out grimly, 'they wouldn't have got out at all.'

They had come to a stop in front of a sub-divided cage. Three hens were housed in one side and four hens in the other – incontrovertible evidence of overcrowding which the farmer could not talk his way out of. Without further argument he removed one hen from each side and put them in a less crowded cage.

Sid Jenkins decided the time had come to caution the farmer and his mother formally. 'Now, I want you both to listen carefully to what I say. You are not obliged to say anything unless you wish to do so, but what you say may be put into writing and given in evidence.' Both declined to say anything, though the woman muttered something to the effect that they had been trying to keep going with one line until they were over their financial difficulties. Clearly the couple had been trying to run two businesses, and finding it all too much for them.

Sid had taken out his notebook once again. 'I put it to you,' he went on, 'that you have many dead birds in the unit and you have failed to remove them. They should have been cleared out and the whole place sterilised. Do you agree?' There was a short pause, then they both replied 'Yes'. Sid continued, 'The birds have been laying eggs for you to sell and you can't even be bothered to look after them properly, can you?' Neither replied.

'What about all the dead birds then?' asked Sid, raising his voice.

It was the farmer who answered. 'That was an illness, it has nothing to do with these birds.'

'So it was a disease?'

'It was egg drop syndrome.'

'What do you understand that to mean?'

'Well, I think it is a cross between fowl pest and various things . . .'

At once alarm bells started ringing. The last thing Sid Jenkins and his colleagues wanted on their hands was fowl pest. The disease is so virulent that even a sparrow entering an infected property can pick the virus up and take it to another farm. Earlier that evening Sid had tried, without success, to contact the Ministry of Agriculture, Fisheries and Food to discuss the conditions at the poultry farm, since it is the MAFF's responsibility to keep regular checks on such establishments to ensure that the birds are being properly cared for. This last disclosure made it imperative that their veterinary surgeons were brought down to the farm in the morning to see the squalid state of affairs for themselves, and to check whether or not the birds had fowl pest.

Meanwhile, Sid was faced with the task of arranging for the birds to be removed from the unit to clean conditions. They could at least try and save these, if not the others that had died already. The farmer suggested rigging up an enclosure and free-ranging them outside, but was reluctant to commit himself to a time when he would do it the next day.

'Well, what time can you be here in the morning?' demanded Sid.

'I'm not sure, I'm under great pressure at our other place,' replied the farmer.

This last remark angered Sid. 'You'll be under a damn sight more pressure if you are not here in the morning. I must point out to you that you are in a very serious position here.'

'Wait a minute,' interjected the woman. 'Couldn't you give us until the afternoon?'

'No,' Sid answered firmly, 'because I am not prepared to condone suffering. I am cautioning you for causing suffering, and you want me to leave things longer than necessary. That would make me party to that suffering. You *will* be here.'

The mother turned to her son and suggested their only course of action would be to send the birds off for slaughter. 'Can you give us until eleven o'clock tomorrow?' she asked. Sid agreed. He would return the following morning with the MAFF vets and would decide then what to do with the hens.

By now it was well after 9 pm. The whole party – the three RSPCA inspectors, the two policemen and John Saxton, the vet, had been at the unit for nearly two and a half hours, and had been questioning the owners for over an hour. As they were about to leave, Sid asked the farmer about the unit's lighting cycle. 'What time will the lighting come on again?' he enquired.

'I think it's eight o'clock in the morning, but I can bypass it,' answered the farmer.

'Will you go and bypass it now so we can see better?' ordered Sid.

'No, I can't do that now because all the bulbs have gone: it won't get any brighter.'

It was pointed out that the lighting was totally inadequate. Not only was the farmer failing to provide the necessary twenty-four hour cycle, there simply wasn't enough light at any time to allow the birds to eat and drink properly. During the day the only extra light getting into the place was through the holes in the roof. 'But they *do* survive,' said the farmer as if in justification.

It had been a long day. Sid's patience had run out. He turned to them both. 'Let's not talk rot. Let's play fair, because we know our job. You've made some mistakes and you're in trouble. We've got to try and help. I'm not vindictive, but at the same time you must understand, if you're not here for eleven o'clock you will be reported for failing to provide care and attention to certain birds and also failing to keep the intensive unit in a proper condition.' With that he closed his notebook and they all left the unit. Sid drove wearily home and had a hot bath.

The following morning Sid was at the RSPCA Group Communication Centre early. This time he succeeded in getting back to the Divisional Veterinary Officer at the

Ministry office in Leeds. After listening to Sid's description of conditions at the poultry farm, he quickly agreed that the Ministry's vets should make an immediate inspection. They would be there at eleven o'clock. Sid spoke next to his boss, Superintendent Len Flint, in York, and arranged for him to be present and assist at the inspection.

No sooner had he put the telephone down than the radio telephone, on top of a filing cabinet in the corner of Sid's crowded office, burst into life.

'Lima Three to Lima Base, over.' It was Inspector John Oxley's call sign. Elaine, the duty co-ordinator, answered it. 'Go ahead, over.'

'I've just called at the farm on my way to join you, and they've been having a massive clean up, so the Ministry vets will not be able to see the situation as it was yesterday. Over.'

The message was acknowledged. So that was why the owners had tried to make the visit as late as possible. They wanted time to clear up the mess before the Ministry inspection.

It was a considerable body of men that made its way to the farm at exactly eleven o'clock that morning. Accompanying Sid Jenkins and Superintendent Flint were Inspectors Manning, Oxley, Wilkinson and a local vet, John Saxton. They were joined by PC Leighton Morris, in a fresh-smelling uniform, and two Ministry vets, a man and a woman, who were wearing large rubber aprons. The farmer's mother was not present. Her son explained that she had other important business matters to attend to. In fact, she was looking after a shop in the centre of Leeds which sold, among other things, 'free range eggs'.

The sun was shining brightly as the party walked down through the farmyard towards the units. It was obvious that a hurried attempt had been made to clear the place up. Five goats were roaming free and one, together with a dog, was chewing at a pile of boxes and sacks heaped up outside the door of the hen shed. A close inspection of the boxes revealed the carcases of many hens. John Saxton, the vet, had with him the report of a post mortem he had done on one of the dead hens the previous evening. It stated that the bird had died from 'chronic egg peritonitis', and that unnecessary suffering had been caused through lack of veterinary treatment. He said that the live bird he had taken away for examination was underweight and had also been neglected. He was prepared to support any action the RSPCA might take in the courts.

Sid and the Ministry vets completed their tour of inspection of the units. He turned to the farmer and asked when the birds were going off for slaughter.

'Well, I can't get a driver. No one will come today,' he replied.

Superintendent Flint asked one of the vets from the Ministry, the woman, what her view was on leaving the birds a further twenty-four hours in these conditions.

'Well,' she replied, 'unless you can find an alternative contractor, they might have to stay. We would prefer them to go off now but I think there is sufficient food and water for them to remain a further twenty-four hours.'

'What about the Environmental Health Office? Are you going to contact them?' asked Sid.

'Um . . . it's not strictly our province, no. We have tried to contact the local authority but could not get through,' she said.

'Have you ever had cause to visit these premises before as Ministry vets?' Superintendent Flint continued.

'We did a few years back. We only visited that house up there.' She pointed to the remains of a hut that now resembled a heap of timber on Bonfire Night. 'I just came along with a colleague,' she added.

Superintendent Flint produced a copy of the Ministry's *Codes of Recommendation*. He showed it to the farmer. 'Are you aware of these codes? They cover advice on how to look after domestic fowl. If you fall below the minimum requirements and standards, then you are heading for disaster. Looking at your birds that are still alive, I would say you have fallen below virtually every paragraph listed in this Ministry document.' The farmer looked bemused and mumbled to himself that he thought the birds were quite sound.

Superintendent Flint went on, 'I would be inclined to say that some of your cages, particularly the ones containing six birds, are over-stocked. You're supposed to give four inches per bird, per drinking trough. There we are – look.' He pointed to the pages giving the tables of correct measurements required for each cage. The farmer looked at them with great interest. The group standing round watched in amazement as he digested the facts in the booklet. It was obvious to all of them that he'd never seen the publication before. But it was too late now for him to start to learn good hen husbandry.

'One of the other points referred to in this document is cleanliness,' Len Flint continued. 'Would you say your stockmanship was all it should be?'

'No, it could be a lot higher,' replied the farmer, his head bowed.

'Well, frankly, I can't see that it could get much lower.' There was a look of disgust on Len Flint's face as he closed the Ministry booklet. A moment of silence followed. Then the farmer looked straight into Superintendent Flint's eyes and made one last appeal. 'I am not neglecting the live birds, and the others died from a disease they had when we brought them in here.'

'But surely,' replied Len Flint, 'when the birds started

dying that should have been a warning to you, and you should have sought advice?'

'I'm not arguing,' whispered the farmer. There was not much else he could say.

Superintendent Flint took out his pocket book. 'Undoubtedly there is evidence of neglect. There is sufficient evidence of neglect for us to advise you that you will be reported for an offence under the Agriculture Miscellaneous Provisions Act 1968 and/or offences under the Protection of Animals Act 1911. On that basis you're not obliged to say anything. If you care to, you can make a statement which we'll record, which might ultimately be used in evidence if the case goes to court. Have you no explanation?'

The farmer just shook his head. He turned and went back into the shed to continue his clean-up. A woman in her late twenties, wearing blue jeans, pullover and protective gloves, joined him. Apparently she was a regular helper at the farm from the nearby village. Sid wondered to himself how anyone could have worked in these conditions and not have reported it.

By now the Ministry vets had carried out their inspection and decided there was no evidence of fowl pest. It was up to them now to supervise the removal of the live birds for slaughter. They promised to let Sid know if there was likely to be any delay, but the matter was now out of his hands.

The next day he heard that the hens were still in their prison. They were likely to remain in their cages for another twenty-four hours at least until a lorry from a slaughter house had been fixed up. Sid felt helpless. Over the days that followed, his thoughts kept drifting back to the farm and the suffering of the hens, though he had long ago typed up his official report and posted it on to the RSPCA's legal office in their headquarters at Horsham in Sussex.

They would decide what action to take on the basis of his evidence. Every day he rang the Ministry and got the same reply. 'We are making arrangements for the hens to be removed.' Seven days after his last visit to the farm, Sid was told that the hens were still there. He decided to call in help from Horsham.

The first taste of autumn was in the air as Alistair Mews, the RSPCA's deputy chief veterinary officer, and Kathy Muriel, the assistant legal officer, arrived in Leeds after a 350-mile journey from Sussex, and entered the Group Communication Centre just after ten o'clock on a Monday morning. The other inspectors who had accompanied Sid on his visits to the farm were also present, and they sat around the small tables in the office, arranged very much like a school classroom, to brief the visitors on the situation there. John Oxley showed them the colour photographs he had taken on the first visit to the battery unit. Alistair and Kathy were appalled. But the main topic of discussion was the plight of the birds remaining on the farm, and why the Ministry had not removed them. It was decided to make a further visit to the farm in an effort to force the situation along. A telephone call was made to the Ministry to tell them of their intentions. The effect was instantaneous. Within a quarter of an hour, as the delegation was about to leave Sid's office, a return phone call told them that a lorry would be arriving at 2 pm to collect the live birds. Sid clapped his hands with relief.

Just after the appointed hour, a lorry laden with empty plastic crates arrived outside the poultry sheds to remove the remaining 429 live birds. But for Chief Inspector Jenkins, the problems were not yet over. The black plastic crates, similar to milk crates, but more finely meshed, were stacked in sixteen rows, eight crates high, on the back of the lorry. This is the normal way to transport poultry around the country (much to the concern of many a motorist following behind one of these lorries with its lurching load of scrabbling birds crammed into their cages like sardines in a can). What was alarming Sid was the condition of the crates. The plastic bars on a quarter of them were so badly damaged, or were missing altogether, that the birds placed in them could escape or be trapped. Sid insisted to the driver that on no account must he place hens in such defective crates. He would be committing offences under the Transportation of Animals Act if he did.

Alistair Mews had been busy contacting another veterinary surgeon who would make a 'third party' examination of some of the hens before they went to slaughter, to try and establish why there had been such wholesale death amidst the flock. No one could understand why the Ministry had not carried out these tests during the previous week. He was soon in touch with a vet specialising in domestic fowl in York, some twenty miles away, who agreed to come over to the farm straightaway.

Alistair Mews is a case-hardened vet, with twenty-five years' experience. He has seen unthinkable suffering inflicted on animals in his time. But he had never before seen live animals kept in such disgusting conditions as those on the poultry farm. Although the farmer had been clearing up for a week, little had been done to ease conditions for the birds still imprisoned in their overcrowded cages. Hens had still been allowed to escape and were running around on the floor of the shed. Most unbelievably of all, the lighting had not been improved. In fact, the three 40-watt light bulbs were still the only source of illumination.

By now, Sid Jenkins, helped by John Oxley, had sorted out enough sound crates for the task of moving the lives birds from the sheds to begin. The lorry was backed up to the entrance of the shed and a human chain was formed to carry the live hens from their cages to the lorry. Sid was working beside the farmer at the cages. What he saw now really beggared belief, despite everything that had gone before. The farmer was grabbing two hens at a time by their legs and pulling them through the tiny gap at the bottom of the cages. He still had not learned from his mistakes. Right up to the last minute he was inflicting unnecessary suffering on the birds in his care. 'Steady on, steady on,' shouted Sid, above the loud din of the frightened hens, 'we don't want to kill them just yet.' Sid patiently showed him how to remove his birds correctly.

As the birds were carried out, four at a time, into the bright sunlight of the farmyard, they were passed to John Oxley, standing next to the lorry, who checked them over thoroughly, then handed them up to the driver to pack into the crates on the back of the vehicle. 'Twenty-four . . . twenty-eight . . . thirty-two,' John Oxley counted aloud as the hens were handed to him. After an hour, all the birds, except two, were loaded up. These two had been kept for examination by the third-party vet from York. He put them in his car to take back to his surgery, where he would destroy them and perform a post-mortem. They would certainly have a more comfortable ride to death than their fellows crammed together in the crates on the lorry.

It was four o'clock when the vehicle left the farm. The hens would be taken away to be slaughtered early the following morning at a chicken factory, their carcases to be used in food products.

There was no doubt in Alistair Mews' and Kathy Muriel's minds, as they caught the train from Leeds back to Sussex, that there was major evidence here of neglect and unnecessary suffering of animals. It was Chief Inspector Jenkins' case, so he would be responsible for gathering all the evidence together and submitting it to his legal office in Horsham, where the final decision to prosecute would be made.

During the next two weeks Sid collected statements and reports from the veterinary surgeons and his fellow RSPCA inspectors involved in the case. John Oxley supplied over thirty colour photographs which illustrated graphically the nightmare of that first visit to the farm. They would be invaluable evidence in court. At last Sid sat down to type his own statement. In all, there were eleven pages of evidence, one of the longest statements of his career. Six weeks later headquarters gave him the go-ahead to start court proceedings against the farm owners.

Otley is a small market town that sits astride the tranquil river Wharfe fifteen miles north of Leeds, surrounded by the rolling hills, dotted with the farms and miles of stone walls, that are the trademark of the Yorkshire countryside. Known as Hotten to thousands of television viewers, Otley is used for the market sequences in the series *Emmerdale Farm*. The poultry farm that was to be the centre of attention at Otley Magistrates' Court this cold, snowy February morning might have been a million, not a handful, of miles away from that pleasant spot.

The farmer and his mother parked their car in the car park next to the courtroom. It was 9.15 and they were early. The court did not sit until ten o'clock. Clutching a brown paperbag full of notes, the man helped his elderly mother from the car. On the other side of the car park a white van pulled into the last remaining space. The snow was now a couple of inches deep as the door of the van opened and Chief Inspector Jenkins stepped out in his best uniform and carrying his briefcase. This was case number CV 1908 in Sid Jenkins' file. He had already been to court twice since the turn of the year. In a couple of hours, case CV 1908 would be over and put away to join the pitiful stories that cram Sid Jenkins' office file.

The courtroom was packed with newspaper reporters, members of the chicken lib movement, the RSPCA people involved in the case, and the defendants with their solicitor. The presiding magistrate looked somewhat surprised to see such a large audience before him. The court clerk stood and read out the charges to each of the defendants.

*'Charge One. Causing unnecessary suffering to a domestic fowl by unreasonably failing to provide the said animal with care and attention.*

*'Charge Two. Causing unnecessary distress to a number of poultry by failing to protect the confined birds from draughts.*

*'Charge Three. Causing unnecessary distress to a number of poultry by failing to ensure that cages were maintained.*

*'Charge Four. Causing unnecessary distress to a number of poultry by failing to provide cages of a size to provide sufficient freedom of movement.*

*'Charge Five. Causing unnecessary distress to a number of poultry by failing to provide sufficient light to the birds.*

*'Charge Six. Causing unnecessary distress to a number of poultry by failing to attend to the prompt removal of dead birds.'*

The man and the woman pleaded guilty to all the six charges, then sat down. They remained impassive, their heads bowed, as Anthony Cumming, the solicitor briefed by the RSPCA, described in detail the conditions Sid Jenkins had found on his first visit to the farm in September, and the subsequent visit he had made with the police and a veterinary surgeon. The man was still holding the brown paperbag full of notes while his mother, in a large overcoat and wearing a large hat covering most of her face, clutched at her handbag – the same one she had put the eggs in as she collected them on that first night when she had gone round the unit with Sid.

The solicitor continued. 'So far as the RSPCA is concerned the purpose for which the Society was formed was for the prevention of cruelty to animals and its purpose in bringing the prosecutions is to expose existing cases of animal suffering. It tries to obtain publicity so that other persons concerned with animals, either by having them as domestic pets or using them in connection with food production, are aware of the existence of the standards expected of them and are encouraged to maintain and where necessary improve their own standards of livestock keeping by learning what befalls those who contravene the necessary standards.'

He then turned to the individual summonses. It took nearly ten minutes to read them all out to the court. Included were the reports provided by the two vets: 'The veterinary surgeon Mr Saxton who attended the farm on the first occasion, gave as his opinion in relation to the overall condition of the premises and the conditions in which the birds had been kept, that there had been a total failure to provide even the most basic standards of care and attention. Therefore all the hens had been caused unnecessary stress for a considerable period of time. Another veterinary surgeon, specialising in domestic fowl, who attended the premises one week later, when quite a lot of cleaning up had been done, came to the conclusion from information and photographs supplied and from his own inspection, that the standards maintained were far below those necessary for the good welfare of any birds maintained within them and the circumstances present would without doubt subject the birds to unnecessary distress.'

Twenty minutes later he summed up the RSPCA's case. 'So far as the Prosecution is concerned your Worship, it is submitted that the evidence shows, these hens have been kept in a battery unit which was totally unsuited by its state and condition for the proper keeping of hens and the photographs exhibited to your Worships amply show the neglect into which the premises have been allowed to fall.' He then sat down.

The counsel for the defendants stood and spoke for fifteen minutes during which he said, 'There had been a degree of exaggeration by the RSPCA. The cages were not wonderful but in an ideal world things cannot be kept in 100 per cent working condition.' Describing the condition of the hens he said, 'They were fit and well, bright-eyed and bushy tailed. My client believed what they were doing was all right, they were harassed by the RSPCA.'

The magistrate then adjourned to his chambers to digest all the facts. He returned after ten minutes and asked the farmer and his mother to stand while he delivered the verdict. 'This is probably the worst case of animal abuse we have heard in this court,' said the magistrate. 'We do feel a grave case of serious neglect and suffering has been caused here and therefore the court finds you both guilty of the six charges against you.' He then fined them £75 on each of the six charges, and ordered them to pay costs of £320, a total fine of £1220.

The snow had stopped as the RSPCA men left the courtroom. It was five months since Sid Jenkins had first discovered the appalling horror at the poultry farm. Five months since he had last been able to eat chicken. Hours of work had gone into bringing this case to court. It was all worth it. No more birds would be suffering ever again at that farm. 'Come on, let's go get a cup of tea,' said Sid with a grin to his colleagues, who were as happy as he was to see a successful conclusion to it all.

As they turned away and walked down the road, the farmer came out of the courtroom. He had the brown paperbag over his face.

# ALL IN A DAY'S WORK

A typical day just does not exist in the life of an RSPCA inspector. Each one is always different; it is impossible to predict what calls for help from the public may come in, or when and where. Emergency investigations just have to be fitted round whatever fixed points are already written into the diary.

Unless he has already been called out to a case early, Sid Jenkins tries to get into his office in the converted Victorian mansion in Chapel Allerton that houses the Group Communication Centre by 9 am. The office consists of two rooms. The larger of the two is where the Group co-ordinators take all incoming calls and operate the VHF radio communications which keeps them in touch with Sid and the other inspectors of the Group when they are out and about. On the walls are large ordnance survey maps of the area covered by the Group. Filing cabinets and reference books occupy the space around the room.

The second office is slightly smaller; here Sid Jenkins and his colleagues have their desks. This is where they sift through the complaints received, and on their return from an incident complete any legal documents.

Sid's first task on getting into the office is to tackle the paperwork. Once the incoming mail has been read and sorted, to be dealt with later as necessary, he will normally spend the first hour of the day typing up his statements for the legal department at RSPCA headquarters, or signing report sheets for the five inspectors in his Group. On this particular morning in July, Sid had got in a couple of hours

On the way to buy some eggs. Sid visits the poultry farm for the first time

John Oxley searches one of the derelict battery units

The horror inside the hen units

Peter at Little Creech before being released into the wild

Sid with Peter the fox

Dandy-Leo, the lioness. She has been found a new home in North Yorkshire

Paul Vodden helps the Council workers carry Zara's body from her enclosure. Nick Nyoka looks on

Time finally ran out for Ricky and Zara

A vet examines a puppy at the RSPCA animal home

earlier than usual since he wanted to be away from the office early. The Great Yorkshire Show, the third largest agricultural show in the country, was being held in Harrogate, and that is an immovable feast in the Group's calendar. Not only does the RSPCA always have a stand there, but the area inspectors must be on hand to deal with any emergencies arising from the presence of so many animals in the showring or showjumping arena.

Just as Sid was about to leave for the show, Elaine, the communications co-ordinator on duty that morning, passed him a note. She's just received a call complaining that the owner of a dog was hitting his animal with a stick. Sid read the details she'd noted down, then glanced at his watch. It was just after nine o'clock. The address given was not far from the office in Chapel Allerton. 'I'll go and check this complaint out before going to the show,' he told Elaine.

The place turned out to be a small corner sweetshop of the old-fashioned kind, stacked with dusty jars of sweets against the walls and rows of confectionary displayed on the counter. Serving behind the counter was an Indian woman dressed in a sari.

'Hullo, I'm Chief Inspector Jenkins from the RSPCA, come about your dog.'

The caller had informed Elaine that the shopkeeper's wife had said that her husband had been so angry when their dog messed on the floor that he'd hit it with a stick.

'What do you want?' the woman asked.

'Haven't you done something wrong?' Sid replied.

'No.'

A dog barked loudly from somewhere at the back of the shop. Sid continued, 'He's been beaten, hasn't he, with a stick?'

Before the woman could reply, a man appeared from a room at the rear of the shop. 'Hello sir. I'm Chief Inspector Jenkins from the RSPCA. I've come about your dog. I understand you've hit him with a stick, is that right?' The man, a Sikh, snapped back, 'I didn't hit him, I didn't hit him. I just pulled on his collar and said you're a bloody stupid dog.'

'Where's the stick you used?' asked Sid. The dog was still barking and was obviously the one he was looking for. 'I'll come round to the back.'

Sid walked out of the shop and round to the rear of the building. Just inside the back door a large alsatian dog was chained to a three-foot lead which completely restricted its freedom of movement. The dog could not move off the back porch or reach the yard outside the building. The back of the shop looked out into a cul-de-sac with a row of lock-up garages to one side.

'Sit down. Sit, sit. Good dog.' Sid stroked the alsatian. The shopkeeper appeared. 'What do you use him for? Guard dog?' asked Sid. The man mumbled something. Sid continued. 'And where was it he messed then?'

'It was over here.'

'And so you hit him for it, did you?'

'I haven't hit him,' shouted back the shopkeeper, who had by this time become quite excited and was still trying to restrain the dog by pulling sharply on its chain.

'Where's the stick you used? Is that it?' Sid pointed to a three-foot cane leaning against the wall near the dog. The man said something, but his voice was drowned out by the noise of the dog, who was barking as much as ever. Sid patted the animal. He calmed down a bit but carried on whimpering. Each time the door of the shop opened or closed, and the door bell rang, the dog started to bark loudly again. The shopkeeper's voice rose as his agitation increased. He shouted back over the sound of his dog. 'I paid three, three, four hundred pound.'

'For what?' Sid looked surprised.

'For the dog,' replied the shopkeeper.

'For that dog?' asked Sid, pointing to the animal.

The shopkeeper didn't reply. He kept on protesting, 'I don't want to hit him, do you think I want to hit him?'

'Somebody saw you hit him,' offered Sid.

'They never saw anything, what are they bloody complaining for?'

There was silence for a while between the two men. Even the dog had stopped barking. Sid did not seem to be getting anywhere. He decided to try again. 'Did your wife go into a shop and say 'It shit all over so I hit it?'

'No.'

'If you keep it tied up like that,' Sid went on, 'it's going to do it unless you take it for a walk. Just how often do you exercise it?'

'Er, in the night after half past ten, when people go away.'

The shop bell rang and the dog started to bark once again. 'Good boy, good boy.' Sid stroked the dog. 'If a dog sits like that and is tied up all day then it is going to make a mess. It's chained up now, isn't it?'

'Yes, I know, but if I don't chain him up he goes through into the shop. I don't want to chain him all the time.'

'I should hope not, because it's no life being chained up all the time. He should be having as much freedom as possible. If I find that you hit the dog,' Sid went on, 'there's going to be trouble: I mean that, you'll go to court, do you understand?'

'Well . . . I don't want to hit a good dog,' replied the shopkeeper.

'Yeah, but you told somebody, or else your wife told somebody, that you hit it.' Sid was beginning to raise his voice. 'You told them you hit it because, in your own words, it shit on the floor.'

'Yeah, well it . . .'

'Well, it will do it if it's tied up there,' Sid interrupted. 'You must take it out more. That chain is so tight that it's choking him. Haven't you got a collar for him?'

'Yeah, I got a collar.'

'Let's see the collar, it's not on now, is it?'

The shopkeeper mumbled something about the collar being downstairs. Sid raised his voice over the noisy alsatian. 'Come on, don't tell me lies, you use that chain around his neck when he's fastened up.' The shopkeeper then produced a choke chain. Sid pointed out that it was not much different from the one already on the dog and it, too, would pull into the dog's neck. He wanted to see the dog with a decent fitting collar. He went on, 'And I don't want you to hit it any more.'

'No, I don't want to hit it,' replied the shopkeeper.

'Right, well don't.'

'I don't want to hit it to kill it.'

Sid looked amazed at this last remark. 'I hope not: it was your wife who admitted you hit the dog. If it's tied up like that it's going to mess on the floor. It's your fault it does it because it's not outside, so don't let it happen again, all right?'

The shopkeeper did not reply. Sid Jenkins had made his point and as he turned to walk away he looked down at the dog. The alsatian was the only one who knew the truth, but Sid hoped that his visit to the shop would act as a sufficient warning to the owner. The sight of a uniform can have a powerful effect. All the same, Sid or one of his colleagues would have to keep an eye on the situation for a little while yet to make sure no further harm was done to the dog.

The visit to the sweet shop had delayed Sid longer than he'd expected. Large traffic jams out of Leeds build up on Great Yorkshire Show days. Sid was glad that his local knowledge of the area had taught him the short cuts through the country lanes, and once at the showground, two miles outside Harrogate, he made his way to the RSPCA stand which he and the other inspectors would use as their base. Soon after his arrival, Sid was joined there by John Oxley and Dave Millard.

The Great Yorkshire showground covers an area of 200 square acres. Like most county shows, it has all the usual attractions – side-stalls, large displays of agricultural equipment and scores of other displays that don't necessarily have anything to do with farming or the countryside. Its main attractions, though, are the vast covered areas where the best of Yorkshire's sheep and cattle are on show. Continually, throughout the day, classes of animals are being judged in one of the many showrings, and professional showjumping events are also held in the main arena, attracting some of the finest riders in the country. The daily attendance at the show is in excess of 100,000 and the cars that brings the crowds are shepherded into six vast car parks. It is these which were of special interest to Sid and his colleagues.

It is a fixed rule of the Yorkshire Show that no dogs are allowed into the showground. Many people, though, persist in bringing their dog which then has to be left in the back of the car – this despite the fact that year after year the RSPCA publicises the dangers of leaving dogs in cars. One of the main duties, then, that the inspectors have to carry out is to check for any dog left in a parked car that may be showing signs of distress from dehydration and overheating, as well as their more obvious tasks of keeping an eye on the livestock at the Show, to make sure no animal is being mistreated in the enclosures or the judging ring, and on the handling of the showjumpers inside and outside the arena.

The weather forecast had promised sunshine and no clouds – just the conditions in which the parked cars would become heated up like ovens. Three years ago a dog that had been found suffering from severe heat exhaustion would have died if it hadn't been immersed by one of Sid Jenkins' colleagues in cold water just in time to bring down its high body temperature. Since then the inspectors always make sure there is a bathful of water standing ready in the compound behind the RSPCA stand.

By mid-morning the car parks were nearly full, and without a breeze to circulate cool air into the parked cars, the temperature inside their unshaded metal frames was quickly mounting. Sid had gone over to the paddocks to check the condition of the horse transporters. His radio, connected to the showground police control, suddenly broke silence: 'Control to RSPCA, Message. Over.'

'This is RSPCA. Go ahead. Over.'

'Can you attend in the fifth car park. A police officer has received a report of a Jack Russell in distress inside a car.'

Sid acknowledged the message, then turned and fought his way against the incoming tide of visitors out to car park 5. There he found a uniformed police officer and Inspector John Oxley already surveying the thousands of cars that were parked in the field, a heat haze rising off their tightly packed chrome bodies dully gleaming blue, red, black, green, as they stretched as far as the eye could see into the distance. Although they had the car's number and description it would be like looking for a needle in a haystack.

They had been joined by an AA patrol man on a motorcycle, and the four men spread out and systematically started to check the rows of parked cars. At 10.47 am they found the car. The little Jack Russell inside was panting heavily and was obviously in great distress. John Oxley tried all the doors of the maroon-coloured Marina and, to everyone's relief, found one of them to be open. The dog was quickly lifted out into the fresh air. Had it been left in its oven very much longer it would certainly have been in serious trouble. As the police officer held the exhausted little dog, John Oxley stuck a printed slogan on the windscreen of the car: 'This is no place to leave a dog on a hot day.' A note was also left on the dashboard asking the owners of the vehicle to contact the RSPCA as soon as they returned.

Half an hour later an irate man and woman, alerted by a broadcast police announcement, confronted the inspectors at the RSPCA stand. No one had any right to take their little dog from the car, they stormed angrily at the men. John Oxley interrupted their outraged flow. 'Do you know that it is against the law to leave your dog in a car, in a manner likely to cause unnecessary suffering?'

'No, I didn't,' the woman replied.

John Oxley pointed out that posters had been placed at all the entrances to the showground advising the public against leaving their dog in the car. 'I am going to caution you both,' said John. 'You are not obliged to say anything unless you wish to do so, but what you do say may be taken down in writing and given in evidence against you.'

The man spoke for the first time. 'Can you tell me what you are supposed to do with your dog, then?' he asked.

'You shouldn't bring the dog!' was John's forthright reply.

'What, leave her at home?'

'You have to make arrangements to leave the dog in

someone else's care. The top and bottom of it is quite simple – no dogs are allowed in the showground.'

The woman, obviously the angrier of the two, was still not satisfied and said so. She felt that the Show organisers should do more. That's not the point, Sid said – the dog had been in distress, and it was this matter they were discussing. It was the police who had called them to rescue the little animal. He carried on, 'Your dog may not have been dying when we found it, but it would not have taken long in this heat for it to have been in serious trouble. You only had one window open a couple of inches and there was no through draught.'

John Oxley told the couple they could be committing two offences: 'One for abandoning the dog in the car, and the other for causing unnecessary suffering.' He took down their names and addresses. The man wanted to know what right the inspectors had to enter his car. If the door had been locked, he was told, the RSPCA would have sought help from the police and the AA to force an entry for the sake of the dog's life.

It took another five minutes of quiet but persistent reasoning to calm the couple down. Sid noticed that the couple had not once enquired about the condition of their dog, who had been left to cool down in the compound behind the stand. Luckily the bath of cold water had not been needed this time. Sid warned the couple not to leave the dog in the car again, then took them to collect their pet. 'That is the end of the matter for now,' he said.

The couple were subsequently given a written warning by the RSPCA. Under the Protection of Animals Act they could have been fined a total of £1000 and/or three months' imprisonment (and a criminal record) for abandoning their dog and causing it unnecessary suffering.

Late morning, and the crowds of visitors were thick now, especially round the main arena where the professional showjumping was going on. Many of the famous names in the showjumping world were present, and were gathered in the practice ring to try out a few fences before the main competition. Sid walked round the enclosure, conspicuous in his uniform. Most of the competitors acknowledged him with a nod of their head as they trotted around, limbering up their mounts.

RSPCA inspectors regularly attend showjumping events like these. Their main task is to keep an eye open for the misuse of spurs and whips, and usually their presence is not needed. Seasoned professionals know better than to take their frustration at failing to clear the jumps out on their horses. But sometimes an up-and-coming youngster, desperately keen to succeed in front of the crowds, will be tempted to blame his or her mount when a jump is refused. Over the years Sid has learned what signs to watch out for. A gentle 'reminder' with the whip after a first refusal is allowable, but if the rider begins to get angry and starts punishing the horse after it has refused twice, Sid will not hesitate to take action.

All seemed to be going well for the young woman rider as she entered the ring this sunny July afternoon. She had recently gained a large sponsorship from a major organisation and had been in the winning line-up at a number of leading shows. Sid stood among the large crowd at the edge of the arena to watch as she began her round.

It soon became apparent that either her riding skills were missing today, or the horse was not in a mood to jump – she was having quite a struggle with her mount as she approached the fences. The horse seemed increasingly in charge as they just cleared the first obstacles, and, when he brought a rail down with a trailing back leg, the rider showed her displeasure with a tap of her whip. Still the horse failed to come to order, though, and refused altogether at the next fence. The girl wheeled the horse round to run at the jump again, but after a second refusal the judges' bell rang to eliminate the horse and rider from the competition.

From his position at the side of the arena near the entrance to the collecting ring, Sid could see – despite the applause of the crowd – that the girl was furious. As she passed Sid, a man approached and seemed to remonstrate with her. She rode past him and dismounted, and as she did so she pulled savagely on the bit in the horse's mouth. Regardless of the fact that a number of her fellow competitors were standing around in full view, she took her disappointment out on the animal yet again, this time jerking so hard on the reins that the bit was pulled right out of the horse's mouth. She then walked away, leaving her mount to the groom who had watched her in disbelief.

Sid Jenkins went after her. 'You must not pull on the bit like that, you'll . . .' he started to remonstrate, but his comment was cut short by the angry young woman.

'Well, you want to try and ride the bloody thing,' she retorted abruptly.

'It's not a case of whether or not I want to try riding it – you mustn't do that to any horse,' replied Sid. He could see that he was not going to get anywhere with her as she was at the moment, and decided to give her a cooling-off period before talking to her again.

The horse had settled down in the care of the groom, and after a quick look to see that no damage had been done to the animal, Sid went to find the stipendiary judge responsible, on behalf of the British Showjumping Association, for disciplining riders. He was in the practice ring when Sid approached him and described what had happened when the rider left the arena. Sid suggested that when she'd had a chance to calm down a bit they should both go and have a word with her. It is a serious matter to ill-treat your mount, and the judge could either issue a firm warning regarding her future behaviour, or take further action. If the matter were to be reported it might even lead to a suspension from future competition.

A number of riders, curious to see what would happen, were standing around as the two men approached the young woman. A few of the older competitors would be glad to see her reprimanded – nothing brings the sport into greater disrepute than apparent cruel treatment of the horses. She'd had a chance to cool down, and stood waiting for the two men to come up to her. Sid told her he'd felt obliged to report her treatment of her horse to the stipendiary judge, and asked her for an explanation. She was ready with her excuse: 'When you come out of the ring and your Mum and Dad come straight up and blame you for failing to jump the fences . . .'

Sid interrupted. 'You still shouldn't take it out on the horse.'

'It's something that happened on impulse,' said the young woman.

The judge then warned her not to do anything like it again. The consequences would be more severe next time. She said she was sorry about what had happened, and the two men left it at that. She'd been given her reprimand, and the chances were she would think again before taking her misfortunes out on her horse on another occasion.

Satisfied that he had done what he could to prevent a promising young rider from damaging her career, and the welfare of her horses, Sid made his way to the RSPCA stand. A call had come in over the radio from the Group Communication Centre. A man had telephoned in asking to speak to one of the inspectors about a fox he was keeping in his house. Since all the available inspectors were committed elsewhere, or were at the Show, Elaine had passed the details on there. Sid decided it needed immediate investigation. He handed the two-way radio set over to Dave Millard to continue patrolling the showground to keep an eye out for any further emergencies that might arise, and made his way back to his van. It was already late in the afternoon as he left the ground.

The small council house was situated at the end of a very long road which stretched for about a mile through the centre of the estate and ended in a large area of open waste ground that appeared to be abundant in wildlife. Most of the houses nearby were being renovated, and some were derelict and boarded up. The man who opened the door to Sid was middle-aged and seemed glad to see him. He pointed towards the waste ground and explained that he had found a young fox alive over there. The vixen and the other cubs in the litter had presumably been killed, and he had taken this lone survivor into his house to care for it.

It came as no surprise to Sid to learn that there were foxes in the area. More and more wild foxes have been moving from the countryside into the outskirts of the city over the last thirty years as their natural habitat has been destroyed by the building of motorways and large industrial estates. Indeed, even deer have been seen in some of the built-up areas of the city. Man's domestic waste provides these animals with instant take-away meals, and his dustbins and backyards have become their chief scavenging grounds.

The man invited Sid into the house. The fox, just under two feet long, was sitting on the rug in front of the fireplace. It had been living in a large cardboard box since the man had picked it up eight weeks before, but it was now clearly becoming unmanageable. He had been treating the cub, whom he called Peter, like a pet dog, and Sid could tell from the way he spoke to it that he was clearly devoted to it. He pointed out, as gently as he could, that although it might not seem like it now, the fox was a wild animal and it would not be fair on it to keep it confined as a pet much longer. Sid's advice would be to release it back into the wild as soon as possible. There is an RSPCA Wildlife Unit at Little Creech in Somerset to which Sid could arrange to have the fox transferred, where it could be prepared for its eventual return to the countryside. Meanwhile Sid told the man to go on rearing the fox as he was doing already, feeding it on best-quality dog food, for a few more weeks until it had become a little more mature.

Sid left the man feeling much happier, and promised to keep in touch with him about arrangements for Peter's future. It was 9.30 when he finally reached home at the end of a day that had started before seven o'clock that morning. As he looked back on the crowded events of the last fourteen hours, he reflected that two warnings had been given – one to the owner of the alsatian and one to the young rider – that would, he hoped, save the animals concerned from further injury and abuse, one small dog had been saved from dehydration, and now he had undertaken to restore a young wild animal to the countryside where it belonged. All in all, a good day's work, he concluded.

Over the next few weeks Peter grew fitter and bigger every day – and full of energy. He could not be kept in the house any longer, and the man made a bunker of concrete and wire mesh in the backyard to house him. It was not ideal accommodation, by any means, but it prevented the fox from becoming too tame and dependent so that it would have been difficult to return him to the wild when the time came. One problem was that four dogs were being kept in a pen in the next door garden; every time they picked up Peter's scent they sent up a constant howling, demanding to be let after him. Peter, for his part, must have been under a lot of stress and spent long periods hidden away in his sleeping quarters in the bunker, the floor of which had been concreted to stop him from digging his way out to freedom.

Six weeks after Sid had first visited the house he returned to collect Peter. As he walked up to the backdoor, all the dogs in the neighbourhood started to bark. There was a line of washing hanging across the backyard, drying in the late afternoon sunshine. The man was expecting Sid and was already standing next to the cage. Peter was hiding.

'He doesn't like strangers,' said the man.

'Come on Peter,' shouted Sid above the noise from the barking dogs.

Together they peeled back the concrete and wire-mesh roof as if they were opening a tin of sardines.

'He'll be better off in the wild,' remarked Sid. 'They're not meant to be in backyards.'

The man stepped inside the bunker, which was six feet long and two feet wide. At one end were the sleeping quarters where Peter was hiding.

'Come on darling, come to your father,' called the man as he put his hand, protected by a large asbestos glove, into Peter's den. He grabbed the fox and dragged it clear of the bunker. The dogs next door, having seen their prey, were going frantic, and the man had difficulty holding the struggling fox. Sid put a choke collar and lead round his neck and took him from the man.

'Does he walk on the lead?' asked Sid.

The man just shook his head, too overcome with emotion to speak.

Sid walked the fox on the lead out of the yard and into the street. The sound of barking dogs was coming from every direction, and one dog was standing outside the gate, ready to leap at Peter. Sid managed to shoo it away as he walked Peter to the van. With one quick scoop he lifted the fox into the cage in the back of the van and secured it.

'Can I have a last look?' The man peered into the cage. 'Hello, my love. You are going to be free now and enjoy yourself. I am as upset as you are. Be a good boy.'

Sid patted the man on the back as he turned away. 'You've done a good job. He'll be all right. You can see they're not pets. We will get him looked after and let you know how he gets on.' Sid shut the door of the van. The man thanked him. He was in tears as he walked back to the house.

Peter was taken to the Leeds animal home for a few days while Sid made plans for the long trip to Somerset. Meanwhile Kevin Manning, based at Huddersfield, had two more young foxes to return to the wild, and all three would be moved down to Little Creech together. One of the foxes in Kevin's care had been in captivity for about a year, and had even taken part in a famous television soap opera. While on the set he had turned on his keeper, a policeman, and bitten him so badly that he had caused severe damage to one of his eyes.

The Wildlife Unit at Little Creech takes in and cares for all kinds of wild animals that have been held in captivity, in order to release them back into the wild as soon as is practicable. It is set deep in the Somerset countryside and is a haven for any creature that has fallen on unfortunate times through no fault of its own. One particular feature of its work is dealing with birds that have been affected by oil pollution.

After a six-hour journey, and a dawn start, Sid handed over his three charges to the warden, Colin Seddon, a former RSPCA inspector. He checked them over before placing them in a large enclosed pen to settle down. It was the first time in their lives that the three young foxes had been allowed the freedom of so much space. They tore around the enclosure at high speed, chasing one another like school children playing tig.

Sid watched in pleasure as they enjoyed their taste of freedom. 'The best place for them is back in the wild,' he said, even though he knew there'd be a chance that one or all of them might be killed by foxhounds within a short time of their release. He took one last look at them before returning home. 'It makes you wonder why people want to tear them to bits,' he said, and got into his van.

Four months later the three foxes were successfully returned to the wild at secret locations all round the country.

# FRIDAY THE THIRTEENTH

'Can you give me advice on a very unusual problem?' asked the voice on the telephone. The caller, the headmistress of a large comprehensive school went on, 'One of our students has had a complete change of character. She has altered from a pleasant girl into a moody one, and I believe that the change has come about because she has been visiting a witch. This witch is supposed to have cut up a dog in the presence of children and to be keeping the pieces in her freezer.'

Sid Jenkins was taking this call at his home during one of his spells of telephone-advisory duty which are done on a rota basis by each of the six inspectors attached to the Group Communication Centre in turn. He asked the headmistress to repeat her last statement. It was not that a witch was involved that was bothering Sid – he has frequently come into contact with people who claim to practise the occult arts – but the fact that only a few months earlier he had been called to a cemetery where the dissected corpse of a dog had been found on a gravestone. It looked as if it had been ritually cut up and sacrificed. Was there a connection?

The headmistress went on to tell him that the parents of the girl were very worried about the incident and would be more than willing to co-operate with any enquiries. She gave him their phone number and then Sid put the phone down. On the table beside the telephone was a logbook. Every call that Sid receives has to be noted down, with a brief description of the problem. He wrote in the log: '21.15:

call from headmistress – witch dissecting animals.' It was the 987th call that Sid had taken at home that year.

The first thing to do was to get more information. He quickly rang the number the headmistress had given him and spoke to the girl's father. He told Sid that he and his wife were particularly worried about a symbol their daughter had recently had tattooed on her forearm. He thought it had something to do with the occult, and had been drawn by the witch. He said he had heard that some animals had been involved in a bizarre incident, and told Sid the name of the street where the witch was believed to live. Sid thanked him, and before ringing off, arranged to meet the girl at her home the following day.

This latest piece of information was duly written down in Sid's notebook, and then he picked up the phone again to call Inspectors Kevin Manning and Steve Wilkinson. He asked them to be in the office first thing the following morning. It would be Friday the thirteenth.

Just after 9 am the next day the two officers met for a briefing session with Sid. He told them, 'The reason for the meeting this morning is to discuss a case which involves the misuse of animals in witchcraft. It has come to my attention that a number of children have been visiting a house in the Leeds area, where I understand, from the headmistress of the local school, in the freezer is the cut-up body of a dog.'

The two officers didn't react to Sid's last remark. In their job, they had learned to expect anything. Sid, sitting behind his desk, carried on the briefing in the policeman-like manner he usually adopts. 'I am led to believe it's a puppy that has been dissected and used for witchcraft. A young girl has given certain information to her school teacher which has been passed to us. I think we ought to go and look at the house involved and keep it under observation. The kids are supposed to go there at lunchtime.'

That was quite likely, Kevin Manning remarked, since the teachers' industrial dispute meant that children were being sent home at lunchtime. Sid mentioned the tattoo on the girl's arm and its possible significance. They wondered where the mutilated dog might have come from, and then decided to keep the witch's house under observation over the school lunch-break to see if any children visited her.

The street where the witch lived was in the southern suburbs of Leeds – an area of back-to-back terraced houses built at the turn of the century. Despite the lack of modern comforts, it's a friendly community where neighbours stop to chat to each other, unlike people living in today's high-rise concrete jungles. Sid and Kevin Manning decided that rather than draw attention to themselves by parking in the street in one of their white RSPCA vans, they would observe the house from a saloon car. Inspector Wilkinson, however, was in his van away from the immediate vicinity.

Neither the headmistress nor the girl's father had been able to give Sid the exact number of the house, but finding it turned out to be easier than they thought. As they turned into the street, a brown mongrel dog ran in front of the car. Sid wound down the window and asked a passing boy who the dog belonged to. 'It comes from that house up there, next to the witch's house.' That was all Sid wanted to know.

There were twenty houses standing on each side of the short cobbled street that looked for all the world as if it was the set for a Hovis advertisement. The witch's house was the fourth down from one end. Sid and Kevin parked their car at the other end of the street, about fifty yards away, facing their target. Steve Wilkinson, meanwhile, was set to patrol the neighbouring street in his van, on the look-out for any children who might be heading in the direction of the witch's house. Both vehicles were in radio contact with each other.

At the end of the street, near to the witch's house, a gang of gasboard workmen were slowly excavating a hole in the pavement. Sid and Kevin, anoraks over their uniforms, slid down in their front seats and watched. They had been observing the house for about ten minutes when two street hawkers appeared. Sid turned to Kevin. 'Now we'll know whether she's in or not.'

The hawkers, a man and a woman, got no response from any of the houses as they systematically made their way up the street. The male hawker reached the house they were observing. It had a green door with a brass knocker in the shape of a running fox. The man knocked on the door and waited. In the car the two men waited, too. There was no reply, and he went on his way. Sid mumbled under his breath, 'Don't try very hard – they just knock once. She's not replying.'

The house was a one-up, one-down, with white-painted window frames. Even though it was the middle of the day, the curtains of both rooms were drawn, giving the house a sinister appearance. Kevin remarked, 'Well, it's obvious why the curtains are drawn, she doesn't want to see anybody, or anyone to see her.'

Alongside every fourth house in the row was a door leading to a small yard containing the lavatories and dustbins for the houses. Steve Wilkinson radioed through just then to say that all was still quiet. Kevin Manning suggested a look in the dustbins to see if there was anything that might interest them. Sid agreed. Inside the dustbin they were interested in Kevin Manning found a heap of sawdust, a note from a pet shop, a scrap of paper with some writing on it, and a page of a newspaper which had had a story cut out of it. Back inside the car, the two RSPCA inspectors scrutinised the evidence. The note from the pet shop was a receipt for two rats, while the scrap of paper contained a shopping list of items, including 'cat repellent and cat wormer', which seemed a curious combination under any circumstances. The newspaper was dated. 'We can trace that, and get a back copy,' said Sid. 'It'll be interesting to find out what the story was and it might give a lead as to what is going on.'

'Mixed in with the sawdust at the bottom of the dustbin were about a dozen empty crisp packets and two shandy tins,' said Kevin.

'That could tie up with the kids,' suggested Sid.

It was now 12.30 pm and no children had been seen near the street. Steve had reported seeing a few groups of children hanging around outside a local fish and chip shop, so it was obvious that the school lunch-break had begun. An hour went by. The only movement in the street had been a curtain in the house to the left of the car. The gasboard men had not moved for two hours. Kevin remarked, 'Is it an extended coffee break, or just their usual lunchtime?'

By 1.30 pm things were still quiet, and the two men were getting peckish. Sid radioed to Steve to fetch some chips from the fish shop, and then he suggested a drive to the *Yorkshire Post* office in the centre of Leeds to obtain a back number of the newspaper. The children could be seen returning to school as they crossed the end of the street, so it looked as though there would not be any visitors to the house that lunchtime.

Half-an-hour later Steve returned from the city centre, only three miles away. He parked his white van round the corner and walked up to the saloon car with the newspaper.

Sid Jenkins quickly turned to page fourteen. They were lucky. The missing story was about a young artist, and the address given was the very house they were watching. Accompanying the article was a photograph of a young woman with long black hair. Sid was delighted. 'We now know her name and what she looks like,' he said.

By now it was mid-afternoon, so Sid decided to send Kevin and Steve back to their respective areas to carry on with their routine calls. He would remain in the street until it was time to keep his appointment with the schoolgirl and her parents. At the end of an uneventful hour, Sid drove the mile or so to the tidy housing estate where the family lived. He introduced himself and was invited into the house by the father. The girl had not yet arrived home from school. 'Can you tell me your side of the story?' Sid asked him.

He sat down in an easy chair next to the fire. 'We were a bit concerned when we heard from the school about my daughter, her behaviour has changed, you know. The teacher said they had heard some girls talking about this lady the children visited at lunchtimes. They said she slept in a coffin and apparently had done something with a pig or a dog.'

Sid interrupted. 'Was it put in a fridge or freezer?'

The father replied, 'Yeah . . . something like that, after it had been cut up.'

Sid asked if he had spoken to his daughter about the matter, and he replied that she had denied going to the house, although she knew the woman. She had told him that other children from the school had been in the house and this is where they had got the story about the 'cutting up'.

'What about the tattoo mark your daughter has?' asked Sid.

'She says she did that herself with a needle,' replied the man.

'Is it deep?'

'There's one on her stomach and one on her forearm. They've been there quite a few weeks now and I er . . . played hell with her when I saw them,' he said. 'It's supposed to be some cult sign.'

At that moment the front door opened and in came the daughter. She looked somewhat surprised to see Sid Jenkins sitting there. 'This is Chief Inspector Jenkins, from the RSPCA,' her father introduced Sid. The girl was in her early teens, and lightly built. Sid asked her if she had spoken to the teachers at school about visiting a witch. The girl said she had, but denied having been inside the house. Sid then asked her if any of the other girls at the school had been inside the house and seen any animals there. The young girl paused for a moment, and then answered that there was a pig, as well as tarantulas and snakes in the house.

Sid interrupted. 'And rats – cats? I'm just asking what you've heard because it's important – some animals may have been dissected. I've had to deal with lots of things like this, that have been cut up, and put on gravestones. We've got to stop this sort of thing happening.' He asked her if she had ever been near the house at lunchtime. Again the girl denied she had.

The father told his daughter to roll up her sleeve and show Sid the symbol on her arm. In the shape of a five-pointed star, it was on her right forearm, four inches above her wrist. It had obviously not been put on by a professional tattooist, and the colours were beginning to fade. Sid asked the girl who had done the tattoo. She insisted that she had done it herself. The girl was obviously upset and worried, and Sid felt there was nothing to be gained by pressing her for more information. If she later thought of anything she'd like to tell him, she should ask her parents to get in touch with him.

It was getting dark as Sid left the house. Back in the cobbled street the gasboard men were still working in the hole they had dug. This time Sid parked the car directly opposite the house he was watching. Lights were on in the other houses, but the one with the green door remained in darkness. He waited. It was now 6.30 pm.

One hour later a small car pulled up outside the house, and a woman got out. She matched the photograph that had appeared in the newspaper. The car drove off, leaving the woman about to place her key in the lock on the door. Before she had had a chance to turn it, Sid had got out of his car and crossed the road. 'Hello, I'm Chief Inspector Jenkins of the RSPCA. I wonder if I might have a word with you about some of the animals you have, and some other matters?'

'Certainly,' the woman replied. 'Come and look at my menagerie. Fourteen to be exact. Did someone tell you I had been cruel to them?'

Sid followed her into the house. Her long black hair flowed halfway down her back and she was much younger than the normal idea of a 'witch'. She seemed almost to expect his visit. There were reptiles and animals everywhere inside the dimly-lit room, in cages and tanks ranged around the walls. In one corner stood a bookcase containing about thirty publications on witchcraft and the occult. A white plastic model of Pegasus, wings outspread, was suspended from the ceiling. A record sleeve next to a record player on a small table featured the picture of a naked woman with a very large python snake draped around her. The room smelt strongly of joss sticks.

Rather than 'go in heavy', Sid decided to let the woman do the talking in the first instance. She began introducing her animals one by one. The first creature Sid met was living in a glass tank in the far corner of the room He was a six-foot long python named Little Lurch, and was fast beginning to outgrow the aquarium that was his home, according to his informative owner. In another tank nearby were Fester and It – two baby pythons, each a mere three-foot in length.

By now the woman, who appeared to be very relaxed, had told Sid that her name was Morticia. The eccentric tour around the room continued. In another tank, two female hamsters played on a plastic wheel, their names Ozzy and Plib. A cage on a shelf behind them contained two very large brown rats whom Morticia proudly introduced as Plague and Pestilence. It came as no surprise to Sid when he was told that the name of yet another large rat nearby was Destruction. Two more rats, a male and a female named Blog and Bleb, completed the rat family. A large rabbit was munching away quite happily on the other side of the room. He had only one eye, so naturally was called Cyclops. Two chinchillas brought an end to the collection of confined animals – or so Sid thought until the woman produced a large hairy tarantula spider from a box on the mantelpiece. She held it in the palm of her right hand and kissed it, then offered it to Sid to hold: 'This is Tish, short for Morticia.'

Four well-fed cats asleep on chairs around the room were more conventional occupants.

Sid had let the woman talk for a good ten minutes by now. It was time to broach the serious aspect of his visit. 'Have you had any schoolgirls visiting you here?' he asked.

'Oh . . . I've had school kids – aye. I've been cruel to one or two because they have been silly,' came the unexpected reply.

'Why, what happened?'

'Well, I had some trouble with them. As you can see I've got a snake and a tarantula, and some nice rats. The kids think, Oh . . . she's got a snake, we can go and look. I've refused some of them because they stole a knife off me, and put a banger through my letter box.'

'What about schoolgirls from the school across the fields? Have they visited you?'

'Yes . . . there were four lasses who came around yelling "Witchy woman – witchy woman".'

Sid now asked the question he had been holding back since he first stepped inside the house. 'Have you got a freezer containing a dead animal?'

She answered at once. 'No. I once did an autopsy on a dog that had been dead for at least six hours though,' she said.

'Are you qualified in any way to carry out an autopsy?'

'No, it's because I know about anatomy, with being an artist,' she replied.

'So you are telling me that the dog that you cut up was dead, and the owners gave you permission to do it?' asked Sid.

'Yes,' came the firm answer.

'Did any of the children see the animal being cut up, or see it after you had dissected it?' he queried.

'No.'

'You see – I was told you had a dog in a fridge.'

'Oh, aye . . . I did keep it in a fridge until the owners wanted it disposing of,' was the somewhat forthright reply.

'Did you use the animal for any sort of ceremony?'

'I have no time for that sort of thing. No.'

Sid then changed his line of questioning. He turned the conversation back to the visits by schoolchildren. 'Do you know a young girl with a star, something similar to this one, drawn on her body?' he asked, pointing to a five-pointed star, made of brass, placed on the table top next to the record player.

'I don't know her. She probably plays about with a ouija board.'

'Are you into anything of that sort?'

'I am into the occult, but only in a small way and I wouldn't do anything to hurt anybody.'

Sid walked over to the bookshelf. He wondered what she meant by 'being into it in a small way'. He turned back to her and asked if she was a member of a witches' coven. She replied that she was most certainly not, adding, 'I practise on my own mainly, either for the benefit of myself, or this lot.' She waved her arms in the direction of the animals – 'Just simple spells.'

They had left the room and were standing at the bottom of the stairs. She clearly felt the interview was at an end. But there was one question yet to ask – something that had been at the back of Sid's mind all the time he'd been in the house.

'I've heard you sleep in a coffin.' He laughed, expecting a total denial. On the contrary; Morticia smiled in agreement and said, 'Yes, I do . . . come on up and I'll show you.'

As she turned and made her way up the narrow staircase, the intrepid inspector followed, stepping fearlessly into the unknown. They entered a small bedroom. The aroma of joss sticks was even more powerful than it had been downstairs. A small red lamp on the ceiling cast an eerie light over the room. Along one wall was a single bed and opposite it a dressing table which had a sign where the mirror should have been which said 'Leeds City Mortuary'. There was a large hammer on it, and a crystal ball. On the wall above the dressing table hung a print of a supernatural figure, with blood pouring from its mouth. And there, lying on a table under the window was a mahogany coffin – complete with brass handles. On the lid was a brass plate, with the inscription, 'Morticia Crowley 1465 to 1490.'

Sid gazed in astonishment at the items in the room. Here he was, on Friday the thirteenth, in a bedroom with a real live witch and a coffin. This was most certainly something that was not included in the RSPCA inspectors' training course. Sid pointed to the inscription on the coffin. 'That is a rather a short lifespan, isn't it? Just 25 years!'

She laughed. 'I'll open it up for you. It's not the same as a conventional coffin, because I sleep in it.' She opened the lid and continued, 'I made this myself, with a little help from a joiner.'

The lifting of the lid revealed a blanket and pillow inside, dented and creased as if someone had been lying on them.

'Are you going to be buried in this?' asked Sid, tongue very firmly in cheek.

'Oh yeah . . .' was the reply, '. . . but hopefully not for a heck of a long time.'

'Not just a 25-year span as the plate on the coffin indicates?' asked Sid.

She laughed and replied, 'Oh no . . . By the way, the police have seen this, you know.'

An object glinting in the dark room caught Sid's eye. It was a large sword. He picked it up. It was heavier than it looked. 'What do you do with this?'

Morticia laughed again. 'It's the sword I use to chase the kids with.'

'I don't think I would like to meet you, charging down the street with that in the middle of the night: you could do someone a serious injury.'

She smiled and said, 'I wouldn't hurt them really. I don't do anything evil.'

Sid stood up straight and took a deep breath. 'Can you

suggest why a young girl of fourteen should be tattooed with a symbol of a three-pointed star on her body?'

'It's something they've done on their own.' She was adamant on the point, and said that the children had got the idea out of a Dennis Wheatley book, not from her. 'It's probably the bad reputations witches have, dancing about in graveyards with all their clothes off.'

Sid felt that he was unlikely to get any more information out of her. The investigation had now been in operation for over ten hours. He was trying to work out in his mind just what had been achieved – if anything. The animals in the house were all in good condition, and there was no lack of food for them. Morticia certainly appeared to care about them. So far as the other allegations went, she *did* sleep in a coffin, but there is no law against that, and even if there were, it was a matter for the police rather than Sid. She had admitted that she'd cut up a dog, but so far as Sid could see no offence had been committed under the animal acts that he dealt with.

He repeated the advice he had already given the woman on the dangers of letting young children witness such activities, because it could have a lasting and serious effect on them. There was no more he could do for the present. He let it go at that, and left the bedroom and went downstairs. He turned to have one last look at the animals in their cages, opened the door, and walked out of the house, without looking back.

Sid returned to his office to find the doors locked – he had left his keys in his van at home. Never mind, he thought – in a few hours Friday the thirteenth will be over.

# SHEEP

RSPCA Inspector Dave Millard was sitting in his van outside Harry Ramsden's fish shop in Guiseley, where the urban sprawl of Leeds and Bradford gives way to the moors of Yorkshire. As usual the queue for fish and chips stretched right round the front of the building – a phenomenon you can expect to see every day of the week during the summer months because this is no ordinary fish shop but 'the biggest fish shop in the world': the restaurant is even adorned with elegant chandeliers and the customers dine at tables laid with the best silver service. The smell of cooking fish wafted through the open window of Dave's van and he was just pondering whether to go and get a bag of chips when there was a knock on the roof of his van.

'Now then, what have you got?' It was Sid Jenkins. He had just returned a swan to a country park lake. This swan had been found some weeks earlier, injured and blackened, in an opencast coal mine. The RSPCA had cleaned and cared for the bird until it was fit enough to be returned to the wild. As he was driving along, he had overheard a radio message sending Dave Millard on a job. 'It's some dead sheep in a field up there.' Dave pointed to some high ground rising up behind the local psychiatric hospital. 'Right, I'll follow you,' said Sid, walking back to his van.

It was late afternoon, but the September sun was still warm as the two men parked their vans and climbed over a stile into the field. The field was large, about the size of two football pitches. Three sides were dry-stone walled – about three feet high – and on the fourth side was a broken-down wooden fence.

'It's supposed to be near a wall, isn't it?' asked Sid. 'Yes, somewhere over there,' replied Dave as the two of them made their way along the stone wall up one edge of the field.

'Oh boy! I can smell it now.' Sid drew his hand under his nose. 'Look at the flies on the poor thing, it's been here quite a while.' The sheep was lying on its side. The fleece was matted, and hunks of fleece were coming away from the flesh in the wind.

'It hasn't been here that long because its eyeballs are still in,' said Dave as he squatted over the decomposing carcase. It only takes a short period of time before the crows peck out the eyes of a dead sheep.

'Poor old thing, it must have been a good animal,' remarked Sid. 'Let's see if there are any markings on it.' Dave lifted up the animal's head for a better look. There was a red idtentification mark on the shoulder. Sid took out his notebook from the breast pocket of his uniform shirt. Every marking that the two men found would be noted. Each farmer marks his sheep with a special identifying mark, so that if an animal wanders away, or is stolen, the owner can be traced and prove ownership.

'Red mark on left back, 16.15.' Sid said aloud as he wrote the description and time in his book. 'Red mark on horn and red mark on left shoulder.' Dave lifted the bloated carcase to check for any further identification but there wasn't any. As he put the animal down a swarm of flies buzzed up into his face.

'Smells nice, doesn't it?' said Sid with a wry smile. He moved some ten feet away from the sheep as he continued making notes in his notebook. Their examination of the dead sheep completed, the two men decided to search the rest of the field. About two hundred yards away they came across a patch of flattened grass. 'Obviously another one's died here and it's been pulled across to the wall over there,' said Sid. He pointed to a shape lying against the stone wall which was certainly another dead sheep. 'So whoever the farmer is, he knows about it. Yet he's failed to bury the carcase, and if any dogs get out here they're going to start pulling it to bits.'

A farmer must be given 'reasonable time' to bury fallen stock on his land. However, it's a requirement of law to bury dead animals, and any person who knowingly and without reasonable cause permits the carcase of any dead sheep belonging to him, or under his control, to remain unburied in a field where dogs can gain access is liable to a fine.

The two men started walking across the field. 'Bloody hell, it's warm,' exclaimed Sid, taking off his cap to wipe his forehead. Just then Dave stopped and pointed. He had seen another sheep. This one was alive, but it was limping badly. They started towards it. 'It's not going to get far,' said Sid as the sheep tried to get away from them. Sid put his cap back on. 'Can you get round the back of it?' he shouted to Dave.

Dave Millard took a short run and a dive at the sheep, but missed. It ran further up the field towards a wall. Sid walked slowly towards it, both arms outstretched. Suddenly his foot caught on something. Looking down, he saw it was a coil of barbed wire. 'Look at that – in a field full of sheep. It's scandalous.' He pulled the barbed wire over to the wall. By now, the sheep had tired of its run, and was hiding in a large patch of nettles. Both men made a dive for it. This time the sheep did not move. They grabbed it by the horns and lifted it out of the nettles, turning it on its back. They bent down to examine it. Dave sighed. 'Look at its feet: they're rotten.'

'They are not so brilliant,' agreed Sid, feeling the animal's thighs. 'Poor thing, we'll have to get the vet up to give it some treatment.' The examination over, they lifted the sheep back onto its feet and let it go. 'Shame,' muttered Sid, as the animal limped away. The sheep was walking on tip-toe like a hunched-back ballerina. 'There's definitely something the matter with its back end,' said Dave, watching it, 'it can't have got like that overnight.'

Sid looked up. A movement on the far side of the field had caught his eye. It was another sheep. 'Look – that one is lame too.' They made their way towards it. It looked in worse shape than the sheep they had just released. This one could only just hobble along. 'That one's lame all right!' exclaimed Sid. 'It doesn't know which leg to limp on.'

As the sheep dragged itself off, half standing, half lying, Sid stumbled across another carcase in the middle of the field. 'Have you seen this one, Dave?'

'That's well gone, isn't it?' he replied. It was little more than a bundle of tattered wool and bones. 'What worries me is they may be carrying a disease, and if another animal picks it up it could be spread to someone else's farm. Let's see if we can find anything else in the field,' said Sid, making further notes in his book. Only a short distance away they came across a sheep's head: bits of wool were still sticking to it. But, more important, it had red paint on one of its horns. Dave logged this latest piece of evidence in his notebook.

By now, they had done a complete search of the large field and accounted for all its occupants, dead and alive. Sid turned to Dave. 'Right, we've got two acres, eight sheep, four of 'em lame and four of 'em dead. But what we've got as well is that the carcases are in varying stages of decomposition. The farmer must have known about this. He's had reasonable time, if he's seen them, to bury them,' summarised Sid, finishing off his notes.

Dave folded up his notebook and returned it to his shirt pocket. 'I'll get a vet up to the lame ones and I'll have a word with the farmer about shifting the carcases. If he doesn't get his finger out we'll have a word with the police and the Disease of Animal Inspectors at the Ministry.' They stepped across the remains of the sheep's skull lying in the field, and walked away, back towards their vans, into a glorious gold and red sunset. Dave, grinning ruefully, turned to Sid. 'Can you see any dock leaves? I've been stung all over the bloody place.'

The next day, Inspector Dave Millard, accompanied by a veterinary surgeon, returned to the field. The vet's examination of the four remaining sheep showed that they were 'broken mouth'. This is a term used to describe sheep – usually elderly ewes – who have lost some of their teeth and are therefore unable to eat properly. They are sold off at the end of livestock sales and are very rarely kept long by the farmer before they are sent to slaughter. For their dead companions, the pain had become so great that they had just lain down and given up the struggle. Although he called several times at the house during the day, and made a search for him in the surrounding fields, Dave Millard was unable to contact the farmer. He managed to speak to the farmer's wife, who promised to tell her husband that the RSPCA were looking for him, but the day passed with still no word from the farmer.

In the evening Dave, accompanied this time by Sid Jenkins, made a last visit to the field. As they stood looking around them they could just make out, down the valley where the farm was, a Land-Rover heading up through the fields towards them. 'Perhaps this is him,' said Sid. Dave sighed. 'I hope so. I don't want to be late home for my tea again tonight.' The two men stood patiently, their arms folded, and watched in silence as the vehicle made its way up the track to where they were waiting.

The sun, low in the sky, cast long shadows of the two uniformed men across the field. As the Land-Rover entered the field Dave Millard advanced to meet it. The Land-Rover halted and the passenger's door sprang open. 'Did your wife tell you?' shouted Dave as he continued walking towards the vehicle. 'What about?' called back the farmer, climbing out.

'Sheep up here.'

'She said you'd called.'

'Yeah, there's four dead along the wall, fairly recent, and there are two lame, and two very lame. Their feet are very bad.'

The farmer looked puzzled. 'There are four dead? They haven't been worried or owt?' he asked in a thick Yorkshire accent. 'No, they haven't been worried by dogs,' replied Dave. The farmer looked towards Sid, 'Can't come up every day can you really?' Sid said nothing. This was Dave's patch and his case, so he must do the questioning.

Dave Millard continued. 'I don't know how often you can manage it, but you have to check your stock.'

'Oh, we don't come up more than once a fortnight. You can't cover every bit. It would take you four hours to walk round.'

'Well, when I first came up here on Tuesday there were sheep in here, alive but lame.'

'They get a bit of foot rot at this time of the year,' replied the farmer. After this last remark, Sid Jenkins could keep silent no longer. 'Yeah, but not so they can't walk altogether.'

'Is it foot rot?' asked the farmer. Sid had seen the report from the vet but was not going to give anything away. 'I'm not saying. All I can say is that they're lame and we've got some dead ones.'

Dave took up the questions again. 'We're a bit concerned. Something should have been done. How long have you had them in here?'

'What – this field up here?' The farmer scratched his head.

'Yeah, this one.' Sid joined in again. 'If you've got dead sheep lying about they can spread disease and, on top of that, if they're dead, you are supposed to bury them.'

'Well, we didn't know they were dead.'

Sid had raised his voice in protest, 'I mean, with due respect . . .', when Dave interrupted, 'And there are another two completely decomposed carcases in the middle of the field as well. Come on, we'll go and have a look at them.'

The three men walked up the field towards the wall where the dead sheep were. The Land-Rover, driven by the farmer's son, headed off after the lame animals. Sid asked the farmer if his sheep had any red identification markings on them. 'Well, I'll tell you if it's my mark when we see it.' Sid nodded. 'You can check the marks out in Sheep Book,' the farmer went on. This is a compilation of ear-tag marks and dye marks used to identify the owner of a particular sheep. The farmer turned to Dave Millard, 'Are you walking over here every day like?'

'No, we've tried to contact you down at the farm but you've not been in.'

'And you knew the sheep were just wandering about without coming to get hold of us?' asked the farmer. This last remark upset Sid, although his family and colleagues all agree that he will usually remain calm in the most trying conditions. He quickly turned to the farmer. 'We came to look for you, now don't start getting stroppy, right.'

The men had reached the first carcase by now. There was a long pause as they stood looking down at it. Then the farmer coughed. 'My own marks are KD.'

'K what?' asked Dave.

'KD.'

'It's marked there, isn't it? That's your marking? Red, back and shoulder?'

The farmer hesitated a minute. 'I mean – if we'd seen it here.'

'Well, that's yours then, is it?'

Dave pointed at the sheep. The farmer put his hand to his head and eased his cap before replying. 'Well, I'm not saying it on oath here now, you know, if you're gonna make a big deal out of it.'

'We think it's yours,' said Dave. 'It's got the markings you told me it had, on the shoulder and back.'

The farmer raised his voice. 'But it don't have the KD on it. You see there are several other similar markings. I can name four off-hand that . . .'

Dave Millard interrupted, 'What, that graze next to yours?' The farmer didn't answer. Sid, who had been busily recording the conversation in his notebook, wanted to know when the farmer had last inspected the field.

'Oh, we couldn't have seen this if we'd walked round,' replied the farmer. 'You've obviously got a lot of time, but we don't have much time.'

'Well, you've got to check your stock, haven't you?' asked Sid, pen poised ready to note down the answer.

'Yeah, but if this was laid here, under wall, and you're coming round to gather sheep, honestly how could you see it?'

'You can see the other dead ones easily enough from the middle of the field,' Dave replied. 'We'll go and look at them now, and I'm telling you that you don't have to say anything unless you wish to do so, but what you do say will be taken down and could be given in evidence.'

After the caution, the farmer remained silent until they

reached the first of the other dead sheep. He prodded it with his book, and turned to Dave Millard. 'How did they die?'

'Well, the vet we've had to see them thinks there's some neglect involved.'

'Old people die, don't they?'

Sid Jenkins joined in. 'Yes, but they can get a doctor, they can ask for help. Sheep can't, can they? It's your job as the owner to look after them.'

'Well, I have my living to make, as I said, and dead ones are no good to me.'

'No,' replied Sid, still making notes, 'and neither are the ones that are lame.'

'Ah, well, the lame ones get right.'

'Aye, but they won't if you leave them and don't come and see them,' said Sid as they moved on to examine the third dead animal.

'It looks like they're your sheep, doesn't it?' asked Dave, standing astride the carcase with both hands on hips.

'Well, there's that possibility. They're in my field.'

It was the first significant admission the farmer had made.

Dave continued. 'They're in your field with your red markings.'

The farmer muttered something under his breath and then turned to Dave. 'Who's reported this?'

'The RSPCA never divulges its sources of information. All I can say is this sheep was alive on Monday and yesterday it was dead. That's when we got the call.' It had been a telephone call from a member of the public direct to the RSPCA, informing them that dead sheep had been seen lying in a field, that had started the investigation.

The farmer raised his voice. 'Why the hell didn't they ring us?'

At this point the Land-Rover arrived carrying the lame sheep. Sid walked over to the driver.

'How many have you got in there?'

'Two.'

'Aren't there some more in the field?' asked Sid, somewhat surprised.

Dave looked in the back of the vehicle. 'There were, but they've obviously got out and gone down the valley into the rest of the flock. There were two others lame when we came up here before.'

'Very lame,' added Sid.

The farmer then told them he had bought 160 sheep at the market in Skipton, and half of them were lame. Sid pointed a finger at the farmer. 'If they're lame and you buy them, they are your responsibility and therefore you've got to look after them.'

The farmer looked at Sid. 'Some of these sheep cost £60 apiece.'

Dave had been looking at the sheep in the back of the Land-Rover. 'If they're lame when you drop them off the wagon you should keep them down there at the farm until you've sorted them out.'

'Yeah, but it isn't as easy as that,' replied the farmer.

'It's never easy, no job's easy. It's certainly not thrifty, having them die on you, is it?' asked Dave. 'You lose out,' interrupted Sid.

The farmer went on, still trying to justify himself. 'We buy these sheep as "broken mouth". We buy them cheap to keep a year and then sell to kill.'

'And whatever the wastage it's hard lines, is that what you're saying?' asked Sid.

'No, no. Everybody has a different policy, some people buy their sheep young.'

'So you deliberately run old sheep for their last year,' said

Dave, 'and presumably they need more attention than the younger ones?'

'Well, I suppose you could look at it that way, yeah.'

'I know it's difficult,' continued Dave. 'We appreciate that, but if you're buying "broken mouth" ewes, you know they are older and less sturdy than normal sheep. It's up to you to make sure you can inspect the number of stock you've got, at reasonable periods. Two weeks is not reasonable really.'

Sid Jenkins closed his notebook. It was obvious the sheep belonged to the farmer and this was a case of neglect. The RSPCA would be taking the farmer to court. As they left the farmer to remove his sheep Sid made one last comment. 'The difference is if you go lame you can go and get a doctor. They can't, can they?'

Four months later, at Otley Magistrates' Court, the farmer was convicted under the Agriculture (Miscellaneous Provisions) Act 1968 for 'causing unnecessary pain and distress' to a number of sheep. He was fined £250. He already had a previous conviction for 'causing unnecessary suffering' to cattle. The maximum penalty which could have been imposed for these offences is a fine of £1000 and/or three months' imprisonment.

# US ENGLISH ARE ANIMAL LOVERS

Sid Jenkins was sitting in his office sorting through the pile of complaints that had been telephoned through to the Group Communication Centre that morning. With him was Paul Stilgoe, a trainee inspector on a course at RSPCA headquarters, who was on attachment to Sid's group for five weeks' 'front-line' training.

'This is one we'll have to look at now.' Sid passed Paul a complaint form. He has a knack of sniffing out the serious cases before ever starting an investigation. This one concerned a man who was supposed to be starving and ill-treating his dog.

The rest of the day's programme sorted out, the two men left the office and set out to track it down. The address given was a groundfloor flat on a council estate in the south of Leeds. The flats were in rows, like terrace houses, and only two storeys high. Sid parked his van about fifty yards away and as he and Paul approached the flat they could see that all the windows, front and back, had been broken and were boarded up with plywood. Sid knocked on the door. There was no reply, so he banged hard on the plywood boarding while Paul shouted through the letterbox.

They were making a lot of noise, and eventually they attracted a neighbour, an elderly woman, who said that the owner would not be back until evening. Sid told her who he was and asked if she'd seen the dog. She said that it was normally kept inside the flat and she hadn't seen it for a few days. Sid went back to the door and lifted the flap on the letterbox. He peered inside, but it was too dark to make anything out, so he asked Paul to fetch a torch from the van. Flashing the light down the hall, Sid could see several scraps of paper on the bare floor, but nothing else. There was no sign of a dog. He put his ear to the letterbox but he couldn't hear any sounds of movement from within – nothing to suggest there was either a person or a dog inside.

Another neighbour leant from the window of the flat above. He told Sid that he had reported the poor condition of the dog to the police. Sid said the RSPCA had only received the report that morning, and would do whatever possible to find the man and his dog. 'He's not fit to have a dog,' yelled the neighbour.

Sid and Paul banged hard again on the boarded-up windows and knocked on the door one more time, just for good measure – it was obvious no one was inside. There was nothing more they could do for the moment. Sid would be on telephone advisory service duty that evening and was due to appear in court the next day, so it would be forty-eight hours before they could return to the flat.

The morning was grey and miserable as Sid parked his white van where he had on their first visit two days before. It was just possible to see the door of the flat from there, but a row of lock-up garages blocked the rest of the view. However, it was the closest he could get to the flat, and if there was no reply this time they would just have to sit in the van and wait until the owner of the dog returned, even if it meant staying there all day. Once more Sid and Paul approached the flat and banged on the door. There was still no reply and when Paul put his ear to the letterbox he could hear nothing. Sid knocked hard on the boarded-up windows and shouted. 'We've come to help you with your dog.' Again the only response came from the flat upstairs.

The same neighbour appeared and told Sid that he had seen the man with the dog on the previous evening. 'It collapsed twice when he brought it out of the flat, and it is in a terrible state,' he said.

This news angered Sid. The poor animal could be lying unconscious in the flat. He banged on the door and windows, much harder this time. If an animal was suffering, he was going to have to find it. Nothing would stand in his way. The noise attracted a couple who were walking nearby. They told Sid that they had also seen the dog in a poor state the previous evening.

Sid spoke to all three by-standers. 'I wish someone had telephoned us yesterday. We could have come down and caught him with the dog.'

Paul continued to knock on the door and shout through the letterbox. If the man was inside, he was making no attempt to answer. One of the neighbours advised Sid to enter the flat through one of the broken windows. 'No, we can't do that,' he replied. 'We can't break into people's houses.' Sid turned and looked at the silent flat. He would have loved to tear down the plywood covering the windows and go inside to search for the suffering animal, but the law prohibited him. 'I'll go and call for some police assistance,' he said, and returned to the van to pick up the microphone on his radio.

'Lima Base. This is Lima One. Could you contact the police and ask them to send the community constable who deals with this estate to this address. Tell him we've had a report of a dog in a collapsed state and we need his assistance – it's rather serious. Over.'

Trish Brown was the operator on duty. She acknowledged the message and replied a few minutes later to say an officer was on his way. Paul Stilgoe joined Sid in the van. 'It's just a matter of sitting and waiting,' said Sid. 'We have got a situation here I'm really concerned about. Unfortunately, in this case we can't go breaking down the doors and windows, so we must wait for the police and see what they suggest. We can't leave the dog in there if it is in a collapsed state.'

Before Paul had enrolled on the RSPCA training course he had been working in a supermarket. He could not imagine a greater contrast to what he was doing now. He felt a pit of excitement in his stomach as he settled down in the front seat of the van to watch the front door of the flat. Sid continued to stare at the flat as he chatted to Paul. He wasn't going to take his eyes off the door, even for a second. He asked Paul how he was liking his new job, now that he was spending his first full week in the field.

'I'm really enjoying it,' Paul replied. 'Since I became interested in the job my ideas about it have changed lots of times. I've been taught so many different skills and I'm proud to be doing something worthwhile to help animals.'

Any further conversation Sid was going to have with his trainee inspector was brought abruptly to an end. The door they were watching opened and a man stepped out onto the pathway in front of the flat. Paul Stilgoe was out of the van first, with Sid in close pursuit as they sprinted the fifty yards or so, past the dustbins and the washing lines, to intercept the man. He was slightly built, with a beard. He stood amazed as the two uniformed men accosted him, but he must obviously have been inside the flat when they were banging on the door and windows twenty minutes earlier.

Chief Inspector Jenkins introduced himself and asked the man if he owned a dog. He said that he didn't, and Sid asked him if he had brought one out of the flat the previous evening. Again the man denied having a dog.

'You were seen with a dog that collapsed outside the flat. you are not telling me lies, are you?' asked Sid. The man became irate and began to raise his voice. Some curtains in the neighbouring flats started to move. Sid asked if they could be shown inside the flat.

'I've never had a dog and I'm not telling lies. You come inside and see for yourself,' replied the man. He took a bunch of keys out from his pocket and led the way in.

It was very dark inside the house. The plywood over the broken windows meant that very little light was able to penetrate the hallway or any of the rooms, and only one light bulb, on the ceiling in the living room was working. It was very heavily shaded, so that in the gloomy light it gave off the room had the appearance of a photographic darkroom. It was sparsely furnished, with only one chair and a settee. There were some posters of a religious nature on the wall. Sid Jenkins asked the man once again if he had a dog. Again he denied it.

'Have you ever looked after anyone else's dog?' asked Sid, who was staring at a spot in the middle of the room. There was a dog bowl sitting on the floor amongst a pile of rubbish. The man saw Sid looking at it. 'It's been there since I moved in three years ago,' he replied.

Paul Stilgoe had discovered another dog bowl and a large bone in the hallway cupboard. These, too, the man said, had been there for three years.

Sid pointed to some dog dirt on the floor next to the bowl. 'How long has this been there?' he asked.

Again the same reply. 'Three years.'

The man asked Sid who had been complaining about him. 'I am,' replied Sid Jenkins. 'Your dog was seen in a collapsed state on Monday night. Is it alive or is it dead?'

The man paused for a moment. He was thinking out his answer carefully. 'Yes, it's dead,' he said. Sid took a deep breath. 'You're saying you had a dog, and now it is dead. Where is the body?'

The man ignored the question. He simply stood and stared at Sid, not saying a word.

Speaking as calmly as possible, since he was beginning to get very put out by the man's uncaring attitude, Sid cautioned him. 'You do not have to say anything unless you wish to do so, but what you say may be given in evidence. I have to tell you that you are not under arrest.'

The man said that he wanted the police to call because someone had broken his windows. Sid told him that the police would be arriving at any moment, but it was the dog they were concerned about, not the windows. Did he want to make a statement? The man replied, 'I don't want to commit myself because I never had a dog.'

At that moment a police officer walked in through the door. Sid told him what had happened so far. The police officer, hands behind his back, wandered round the room and then went into the kitchen, situated off the far end of the living room. He came out carrying an empty tin of dog food containing a fork.

'Do you eat dog food?' asked Sid.

'Yes.'

Sid held the tin of dog food close to the man. 'So you eat dog food, do you?' he asked again. Again the man said he did.

'Don't tell lies, you've fed the dog with this.'

Sid turned to Paul and asked him to fetch a black bag out of the van to put the tin in. It could be needed in evidence.

'Well, sir,' said the policeman, turning to the man. 'The position is this. These chaps are from the RSPCA and they are looking into the matter of a dog that has been seen in your care and in a poor condition. If the dog is dead, tell us where the body is so that we can ascertain what was wrong with it. You see us English are animal lovers and these chaps want to help you.'

The man remained silent. 'It would save us a lot of trouble if you would tell us where the dog is,' said Sid. The three of them were standing with the man in the kitchen doorway. Sid was holding one tin of dog food and the policeman another. None of them was expecting the man's reply.

'He's at the RSPCA,' he said.

'When did you take him there?' asked Sid.

'Yesterday.'

'Why didn't you tell me that before? Did you sign the dog over to the RSPCA and get a receipt for it?'

'No,' replied the man.

He told Sid that he could not afford to keep the dog, and that was why he had taken it to the RSPCA animal home. It seemed likely that he'd found out about Sid and Paul's visit two days earlier, and had taken the dog to the animal home then, thinking it would not be traced. Sid could easily check his story out. He returned to the van to radio the control room and within a few minutes received an answer back. A man of that description had taken a dog, in very poor condition, to the animal home in the city the day before. He had said he'd found the dog in the city centre and had handed it in as a stray. He had given his own name, but a different address.

Inspector Dave Millard was on the other end of the radio in the Group Communication Centre, acting as the liaison between Sid and the animal home. He confirmed that the dog was very ill and weak, but was still alive. The animal home will usually keep a stray dog for seven days to allow time for the owner to collect it. If the dog remains unclaimed after that time, the animal is humanely destroyed. Arrangements were now made, through Dave

Millard, to transfer this dog from the stray dog kennels to those set aside for animals that are the subjects of cases currently being investigated by the RSPCA.

Sid Jenkins then returned to the flat. He told the man what Inspector Millard had reported to him on the radio about the dog, but there was still no reaction at all from him. Once again he just stood and stared at Sid. He was cautioned again and asked to give his full name and address. The man refused to do this in front of the policeman. 'The police are not good for me,' he said. Sid asked him why he had given a false address. 'I used to live there,' he replied.

Sid then told him that the dog would be examined by a veterinary surgeon; he could select a vet of his own choosing if he wished. But the man refused to say anything further, and Paul and Sid, together with the policeman, left the flat.

On their way back to the van the two RSPCA men were approached by anxious neighbours wanting to know what had happened to the dog. Sid told them it was being cared for at the RSPCA. Having made sure that the neighbours would give statements if required, Sid set off with Paul Stilgoe for the animal home.

There the staff were waiting for them. Sid walked into the main office and spoke with the manager, Jack Fletcher. Now that the dog had been traced, and was safe, there was a chance that a good home might eventually be found for it. Jack Fletcher showed Sid the form that had been filled in by a member of staff when the dog was handed in. It contained a false address and gave the location in the city centre where the man had supposedly found it straying. One of the kennel assistants said the dog was very thin. Sid asked what kennel it was in.

'Thirty-nine,' said the kennel assistant.

Accompanied by Paul Stilgoe, Sid walked down the passage that led from the office into the kennels. The staff had just finished hosing down the floor and the noise from the inmates was exceptionally loud. Sid approached the kennel where the dog was housed. It was situated in a block of ten that were placed in two rows of five, facing each other. Sid counted off the numbers: '36, 37, 38, 39.' He stopped. What he saw shocked him. 'Oh bloody hell, you poor little sod,' he gasped.

He quickly turned the key to the kennel and went in. Bending down, he stroked the pathetic bundle of bones that was supposed to be a black and white adult dog. The long hair concealed the true condition of the animal. Sid felt the bony prominences of its skull, ribs and pelvic bones. It was extremely weak and underweight. All the same, its large, brown friendly eyes looked appealingly up at Sid. Although suffering, it tried to express its pleasure at being stroked by wagging its tail. It was probably the first time in its wretched life that anyone had shown it so much affection.

'You don't deserve this, do you?' said Sid, visibly moved, hugging the dog. He showed Paul Stilgoe the dull, harsh state of the dog's coat and the sores around the roots of the hair, and got him to feel the dog's bones. As a trainee inspector, Paul was unlikely to see a dog in a worse state than this. Even Sid, after thirteen years in the job, could not remember worse neglect. The dog could not stand up. 'No wonder he keeps falling over,' said Paul. 'He's got no muscles to hold him up.'

Sid went straight back into the office to telephone the vet. He wanted a full examination of the dog as soon as possible. He did not have to wait long. The dog was soon on the examination table in the animal home clinic. 'Terribly thin, in a shocking state, and very anaemic,' said the vet as he carefully probed with gentle hands the dog's slender frame. He picked the dog up and stood it on some bathroom scales. 'It weighs just fourteen pounds. A dog of this age and sort should weigh about thirty,' he said.

Sid noted this information down. 'That means the poor animal is under half the weight it should be,' he remarked wonderingly, shaking his head in disgust.

'That's right,' said the vet. 'The weight loss is very obvious, even to a lay person. The dog looks horrific.'

The animal was placed gently back on the table and the vet continued his examination. 'It obviously hasn't eaten properly for a long time. You'll have to be very careful about how you reintroduce food,' he said to Sid and to the girl assistants who were standing in the room watching the examination. 'He needs feeding gradually. You will have to take him right back to the weaning stage and start him off on liquids, building him up slowly.'

In his opinion, the vet went on, turning to Sid, the dog had not been provided with necessary care and attention over a long period, and had been caused unnecessary suffering. Sid pointed out that two empty tins of dog food had been found in the flat, and so the vet took further blood samples which later established beyond any doubt that the dog's anaemic condition had been caused by lack of food rather than illness. The vet added that he would be prepared to give a statement to support any action in the courts if the RSPCA decided to prosecute.

The vets who visit the animal home daily, together with the staff, tried desperately to improve the condition of the dog over the next few weeks. Further tests were carried out on the dog to find out if there was any underlying medical problem which would prevent him from putting on and maintaining weight. The dog was fed like a young puppy, but despite everyone's best endeavours, he did not seem to improve.

Sid visited him regularly and was disappointed at the dog's lack of progress. Occasionally there appeared to be a glimmer of hope when the dog seemed to gain a little weight. Sadly, however, because of irreparable damage caused to the internal organs through lack of feeding, the dog was put to sleep on veterinary advice six weeks after Sid had started the investigation.

There are some six million dogs in Britain, of which less than half are licensed. The remainder are not licensed because the police have neither the time nor the inclination to seek out the owners, and the work of doing this has not been delegated to anyone else.

Because it is so easy to obtain a dog from any pet shop or dealer, and because half the owners do not bother about a licence, a great many dogs are acquired casually and without any sense of responsibility. There are an estimated 500,000 stray dogs in the UK at any one time and of these, 200,000 are handed to the police each year. Some are accidentally lost, but the majority are dogs which are neglected by their owners and allowed to wander at will for hours and sometimes days on end. Many cause accidents on the roads, resulting in their own injury or death, and in some cases injury or death to humans. A recent report from the Department of Community Medicine at the University of Manchester put the figure for dog-associated road accidents at about 54,000 each year.

In rural areas, stray dogs cause the deaths of thousands of sheep and livestock. According to the National Farmers Union around 10,000 head of livestock are killed or maimed by dogs each year. Most of these dogs are not vaccinated and contribute the spread of serious canine diseases, such as distemper and parvovirus. The mating instinct of stray dogs leads them to harass properly controlled bitches in season, and where bitches are inadequately restrained, many unwanted pregnancies inevitably result.

The RSPCA is adamant that no dog should be allowed outside the home and garden unless it is on a lead or under someone's direct control. On the large council estates in Sid Jenkins's area there are dozens of dogs running loose in every street every day. The RSPCA is convinced that only sensible dog control is likely to persuade irresponsible owners not to allow their dogs to stray. Many of the strays are young, healthy dogs, casually acquired on the spur of the moment and discarded equally casually with complete disregard for their future. Large numbers are taken to the RSPCA at holiday times because their owners will not pay the cost of a fortnight's boarding kennel fee. A considerable number of these have to be destroyed, quite unnecessarily, because there are not enough good new homes to go round. In 1985 the RSPCA alone had to have 52,000 dogs humanely destroyed. This destruction of unwanted dogs is a shameful commentary on the indifference of many so-called animal lovers.

There is now an urgent need for improved legislation. The introduction of commercial puppy dealers and farms has simplified the process of purchasing a puppy, and impulse buying is on the increase. There are, of course, other factors, such as people moving into flats where dogs are not allowed, and more women going out to work, to account for the growing numbers of strays. However, nearly all dogs taken to the RSPCA for rehoming come from those who have obtained them without any proper thought for the future. The situation is likely only to get worse if the government's proposed measure to get rid of the dog licence altogether, rather than raise the fee to a realistic level and administer it properly, is put into effect.

And there will be thousands more unwanted, neglected and half-starved dogs on the streets for Sid and his overworked colleagues throughout the country to try and help – more often than not when it is already too late to save them.

# ZOO CLOSEDOWN

'There's only one thing for it, they'll all have to be put down, and I hope to God I'm not here that day, because you might as well put me down at the same time.'

Nick Nyoka, co-owner of Knaresborough Zoo, was speaking to Chief Inspector Sid Jenkins. His words hung in the air on that cold and miserable November morning. All too shortly, much sooner than either men would have imagined possible, they would be faced with just that situation.

Knaresborough is a charming old-world market town, near to Harrogate, in North Yorkshire. Picturesquely placed astride the River Nidd, the ruins of the castle overlook a steep river gorge. Its quaint market square with its Georgian architecture, including the oldest chemist's shop in England, make it a favourable place for visitors. It is also famous as the birthplace of Mother Shipton, the eighteenth-century prophetess who predicted the end of the world. Few of its citizens can have dreamed then that their little town would soon feature so prominently in the newspapers, both local and national, and receive massive radio and television coverage in the weeks to come.

A year before, in December 1984, Knaresborough Zoo had been inspected on behalf of the Secretary of State for the Environment. His inspectors recommended that a licence for the zoo be refused. This has been required by law since 1981, and has to be renewed annually. The inspectors did not believe that the zoo's owners had sufficient financial backing, or managerial expertise, to bring the zoo up to an acceptable standard of modern zoo management. In March 1985, Harrogate District Council convened a public meeting to consider the licence application in the light of the Secretary of State's recommendations, at which the RSPCA's Chief Wildlife Officer gave evidence against the zoo. The licence was refused, but Nick Nyoka and his wife Barbara, his co-owner, decided to appeal against the decision.

The appeal was heard in Ripon Magistrates' Court between 18th and 21st November 1985. The Chief Wildlife Officer, with other expert witnesses, again gave evidence against the zoo. The court heard of an escaped tiger that had eventually been shot by the police: evidence was given of insecure locks, of fencing of inadequate height, of enclosures that were too small for the animals. Concern was expressed about the numbers of animals which had 'disappeared' over the years. The zoo records were clearly inadequate, although they did reveal that a monkey called 'Dandy', which became ill, was denied veterinary attention. The animal died ten days later.

The appeal was rejected and the owners of Knaresborough Zoo given six months to dispose of the animals, and close its doors. The case is considered by many to be a milestone in the development of modern zoo standards, for had Knaresborough Zoo obtained its licence, the Zoo Licensing Act 1981, which had come into effect in April 1984, would not have been worth the paper it is written on. Its provisions introduced, for the first time, a system for licensing zoos in Great Britain. Before this there was no government department with particular responsibility for overseeing the upkeep and running of zoos; usually classed as 'leisure parks', they came under local authorities, and without any agreed code of standards, particularly with regard to the keeping of wild animals, there was great variation in the way safety regulations and other provisions were enforced. The Act had therefore come about as the result of pressure from organisations who wanted government legislation to provide much stricter regulations.

A zoo licence is granted for a period of four years before it is renewed. The new Act gave the local authority discretion to attach any conditions it thought necessary or desirable to ensure the proper conduct of the zoo during the period of the licence. The Act also stated that a licence would not normally be refused in a situation where adequate standards were not actually being met but where there were reasonable prospects that improvements would take place. It was obviously felt in the case of Knaresborough that this obligation, amongst others, would not be fulfilled.

Sid Jenkins had been stationed in Harrogate as a local inspector before being promoted to Chief Inspector in Leeds. He knew the zoo and its animals very well, recognising each by name, and was acquainted with most of their habits. Each was a character in its own right and Sid could not entertain for a moment any suggestion that they should be put down. After all, they were the innocent victims in a drama that was not of their making. He left the zoo that November day, therefore, with a heavy heart. He had told the Nyokas that he would help in any way he could. He did not know, though, what he would in fact be able to do. As he left, he took one last lingering look at the big cats. They were really his favourites, and he wondered what the future held for them. Just three weeks later he would be back, forced to make decisions that would mean life or death to these very animals.

In its heyday, over 40,000 people had gone clicking through the turnstiles of Knaresborough Zoo each summer. During the last four years, though, debts had been mounting up as attendance figures had dwindled. Mrs Barbara Nyoka had come to terms long before her husband with the possibility that the zoo might not be granted a licence. She had wanted to phase out the larger animals, change the emphasis to that of a Children's Zoo, and look for other attractions to draw the public. Nick Nyoka, meanwhile, continued to devote all his energies to building up his collection of reptiles which was the envy of many: he was considered to be a leading expert in the field. A short and plump figure, about 5′ 5″ in height, with long, dark hair that was tied at the back in a pigtail, he walked with a limp, the result, so some said, of a wrestling match with a lion that didn't quite go his way.

Nick's trips to the Everglades in Florida to restock his collection had no doubt put an extra strain on the zoo's financial position. He seemed to be burying his head in the sand, hoping that the problems would go away. They did not. Once the appeal had gone against the zoo, one of the first creditors to act was the electricity board. The board had threatened on numerous occasions to cut off the zoo's supply of electricity – vital both for lighting and heating many of the cages. Although special arrangements had been made to cope with payments, despite all his efforts, Nick had not been able to meet them. It was at this point that he called in Sid Jenkins.

The panic call reached him on Monday morning, 16th December. At the same time, as if to confirm Nick's worst fears, Sid received a message from the electricity board advising him of their firm intention to disconnect the power supply. He left immediately for the zoo, travelling the fourteen miles – a journey he was to make many times over during the coming winter months – as quickly as he could. Uppermost in his mind was the well-being of the animals, and on arriving at the zoo, he sought out both the directors in order to discuss the situation with them together. This, like many other things at the zoo, was not that easy.

The pressure of events leading to the present crisis had split both family and staff right down the middle. Nick Nyoka was living in a caravan near the entrance to the zoo; his wife and son were living in another caravan on the other side of the site. Husband and wife were not speaking to each other. Sid felt sorry for both sides, but was determined to try and remain neutral, so that he could achieve the best possible results for the welfare of the animals.

Having established that there was no money available to pay the electricity bill, Sid contacted the board in an effort to buy time. Those that would be most at risk if the supply were to be cut were the fish and the reptiles. With the loss of a constant temperature to their tanks, these unfortunate creatures would be dead within hours. When he explained what he was hoping to achieve, much to Sid's relief, the officials at the electricity board proved to be understanding and supportive. They agreed to leave the power on for four more days, until Friday, 20th December, to allow time to persuade the Nyokas to start finding new homes for the animals and reptiles. Now Sid's most difficult and urgent task was to make the Nyokas aware, in separate discussions with each, of how desperately quickly they needed to sort things out. There was now no time to stop and plan. An inescapable deadline had been set.

The Nyokas have lived and worked with animals in British zoos for over twenty years, first on the Isle of Wight, then at Colchester and Sherwood, before coming to Knaresborough some fourteen years ago as curators, later taking over as owners. When Sid Jenkins had first visited the zoo eight years earlier, it had been a ramshackle place, very unsafe for animals and humans alike. Since then the owners had tried very hard to make improvements. What had once been wooden cages and wooden fences were replaced by wood and metal, but even these were now in a state of disrepair and did not meet the requirements of the new Zoo Licensing Act.

Sid asked Nick Nyoka to go around the zoo with him to see how he might be planning to alleviate the animals' plight. Standing outside the enclosure which housed Galan, a mandrill ape, Sid asked if he had made any arrangements for the animals yet.

'No,' replied Nyoka, 'I'm taking each day as it comes.'

'Knaresborough Zoo is now finished,' continued Sid, 'and you have got to accept this.'

'Well, I suppose so. Do you think zoos will disappear altogether?' asked Nick.

'People can see animals and wildlife in their natural surroundings, on their television screens and at the cinema,' suggested Sid.

'It's not the same as a day out at the zoo, though, is it?'

'Surely the best place for the animals and snakes is where they belong, in their natural environment?' Sid went on.

'I'm against dragging animals out of the jungles. All my animals have been bred in captivity. Even Galan, the ape there, was born in Southport Zoo in Lancashire: he was brought up in a caravan,' Nick replied, taking a line that Sid half expected.

Sid agreed that Lancashire could not be classed as a jungle area unless, of course, you are a Yorkshireman. He could not resist pointing out that Nick himself had made frequent trips to the Everglades to bring back rare snakes. Nyoka told him simply that if you had been brought up with animals and lived with them all your life, there *was* no other life. They moved on, leaving the ape gazing at the two men from behind the bars of his cage.

They stopped next in front of the enclosure that housed Roma, the Sumatran tigress, a sleek, elegant animal who came right up to the front of the cage to greet Sid as he approached. She was an old friend. He stroked her and she nuzzled up as close to him as the wire would allow. What would the future hold for this magnificent creature?

With the same feeling of despondency Sid Jenkins contemplated the fate of the other large animals in the zoo – twenty-eight of them altogether. There was Tank the tiger, Satan and Ricky the lions, and Dandy-Leo the lioness who –if you asked her nicely – would roll over and over on the ground for you. Kaffa, a loveable panther or – more correctly – a black leopard – was old and had lost most of his teeth, but could still tackle his food with relish. Less fortunate was Zara, an elderly puma, who was blind and had a large growth: the future looked decidedly bleak for her, as it did for the family of three Himalayan bears living next to her: Yogi, the adult male who was also blind, Dolly and the youngest bear, Treasure, who was only four.

Among the other animals whose future was causing Sid concern were five monkeys (in addition to Galan the ape) and a family of tree bears. There were also four dingoes, a silver fox, a goose, a peacock and many other birds and domestic animals. The problem of finding new homes for all these animals seemed enormous, but it was clear that unless the search was started at once many of these beautiful and friendly animals would be facing certain death in the next few weeks.

Friday, 20th December came, and the electricity board told Nick Nyoka that they were prepared to extend the deadline until after Christmas. The new cut-off date was 3rd January. Christmas came and went, and just after the end of the holiday a new crisis occurred. It appeared that on Christmas Eve one of the dingoes had escaped from the zoo. What particularly bothered Sid, when he heard about it, was that its disappearance had not been reported to the authorities, since dingoes are on the list of animals covered by the Dangerous Wild Animals Act 1976. It is only fair to point out that this particular dingo had lived as a pet with the other domestic animals in Nick Nyoka's caravan and was only put into an enclosure at night. Sid felt he had no option but to report the matter now, and immediately drove to Harrogate Police Station where he spoke to a senior police officer.

At 6.30 am the following day – the day that the electricity supply was due to be disconnected once and for all – Sid left home to drive through the dark to the zoo. As he arrived the sun was beginning to appear over the eastern horizon. He entered the gates of the zoo through a large crowd of press, radio and television men who had gathered there already, although the supply would not be cut off until 9 am. Sid made straight for the reptile house, where he found Nick Nyoka walking up the long viewing area between the glass cages, virtually in tears. From time to time he paused to gaze at his beloved reptiles basking in the heat of their special lamps. As Sid was discussing with him what steps they would take to move the snakes when the electricity supply went off, a woman television reporter thrust a microphone at his face. It was the first of many television interviews that would come from Knaresborough Zoo during the coming months.

'Mr Jenkins – how worried are you about the missing dingo?' she asked.

'I know this particular dingo. It has been treated like a

pet: but, nevertheless, it is classified as a dangerous animal and as such had to be reported,' Sid replied.

Having got her foot in the door, the next question came fast: 'What is your responsibility here this morning, assuming the electricity is cut off?'

'Well,' replied Sid, 'I'm here to assist in every way to make sure that no animal suffers: there are problems here and somehow they have to be resolved.'

Another reporter joined in: 'What do you hope will happen to the animals in the long term?'

Sid thought long and hard before answering this one. He realised that what some of the reporters wanted to hear was the sensational statement that would catch the headlines – the announcement that the animals were all to be put down. The reporter waited patiently for Sid's reply. 'It would be nice if everything could be found a home. Some of the animals, such as the big cats, have little chance of finding a place because they have been brought up as single animals and would not mix with others of their species.'

'What will happen to them, then?' he was asked. To Sid's surprise, Nick Nyoka answered for him. 'They'll have to be put down,' he said.

'That's going to break your heart, isn't it?' asked the same reporter.

'Oh yeah . . .' replied Nick, his head bowed.

'And not just Nick's heart,' said Sid.

It was a good 'out' line. There was a long pause and the interview was finished.

By 9 am, the temperature was well below freezing, and it was snowing as the electricity board officials walked into the zoo to cut off the supply. They went straight into a small building at the back of the monkey house which contained the electricity meters. Within a few minutes, the zoo's life line had been cut. It happened so quickly that it caught Sid by surprise. Immediately the lights went out, and all the heaters in the reptile house began to cool.

Sid was joined at the zoo that morning by David Hornsey, the principal Environmental Health Officer for Harrogate Council. The two men were to work very closely together over the next few months. Now, their first joint action was to confront Nick Nyoka in his caravan where he had retreated as soon as the power was disconnected. The place was full of mementos of bygone days, mainly photographs of a younger Nick sporting safari gear and hugging his beloved big cats. As the three men sat together in the lounge area of the caravan, it was obvious that Nick was in a state of shock. He had thought this moment would never come.

'They couldn't get in there quick enough, could they? What will happen now?' Nick laughed nervously, the stress he was under apparent in the rising pitch of his voice.

It turned out that Nick had made arrangements for the venomous snakes to be collected by another zoo the following morning. 'But I was counting on having electricity to see me through until then,' he added. Beyond that he had no idea what would happen to the animals; nor would he discuss how the electricity arrears might be paid off.

Sid meanwhile had spoken to the two men from the electricity board and discovered that the supply could be restored within one hour of a request for reconnection being made, if an agreement for future payment could be worked out. This made Sid feel a lot happier. He knew that the snakes would last at least twelve hours without heat and, with the use of butane gas heaters, even longer than that. His next step was to ring the RSPCA's Chief Wildlife Officer, Stefan Ormorod, in Horsham. Quickly he filled him in on the latest situation. At the end of their fifteen minutes' conversation Sid, who had been very solemn all morning, started to smile. The RSPCA had agreed, on compassionate grounds, to pay for the future supply of electricity for Knaresborough Zoo until the welfare of all the animals had been secured. This was a very big decision for the RSPCA to make, but one which would win them approval and support from people all across the country.

More discussions followed with the zoo's owners, in the course of which it became clear that Nick's thoughts and energies centred solely on the fate of his snakes. So far as the large animals were concerned, he was leaving it all to Barbara to take on the responsibility of rehousing them. At this stage, it emerged that for some months Nick had been unable to provide adequate food for the big cats; without the help of a local animal sanctuary, their situation might have been far worse. Sid now obtained a letter of authorisation, signed by both Nick and Barbara, entrusting the RSPCA with the job of finding new homes for the animals: if none could be found, it would arrange for them to be humanely destroyed. It was agreed that Nick would continue to try and rehouse his snakes.

Feeling that he was now beginning to get somewhere, Sid went in search of a phone to contact the electricity board and arrange for the reconnection of the supply as soon as possible. Making his way along the muddy pathway between the cages, he passed a television crew recording an item for the next news bulletin. He paused and listened.

'A spokesman for the North Eastern Electricity Board said they regretted today's move to withdraw the zoo's power supply – a decision which was only made as a last resort following numerous meetings with Mr Nyoka who, says the board, has continually failed to make weekly payments on his outstanding bills. RSPCA officials plan to try and get the power restored, but for the animals remaining out in the cold at Knaresborough Zoo the future looks extremely bleak. This is Nick Wood, for BBC Television News at Knaresborough Zoo.'

Until his interview earlier that morning Sid had had no real experience of the news media. However, quick to learn, he had already realised their power to move the public. Here was his chance to make a direct appeal to help save the lives of the animals. Realising the press were after a story, he spoke to the reporter. What the RSPCA was trying to do, he said, was to find new homes for the animals. Those who could not be rehoused would have to be destroyed: their agony couldn't be prolonged any further. His urgent message went out over the air on the lunchtime television news.

Immediately the telephone lines to the zoo and the RSPCA were jammed with calls. Within half-an-hour all the domestic animals – the goats, cats and rabbits – had been offered homes. Sid's interview had paid off. As the weeks went on, he would become ever more confident in handling the press and television reports, maximising their potential to bring in offers of help and money from the public to save the animals.

A tense afternoon followed. Working with David Hornsey, Sid set about trying to secure future homes for all the animals. It was a tricky, often frustrating, task. At one time they managed to arrange transport to remove some of the snakes and Galan, the mandrill ape, to Dudley Zoo. Then Nick changed his mind. He'd got other plans. He was obviously finding the stress of the day's events difficult to cope with. Perhaps if he arranged to sell some of the rare birds and snakes it might ease his immediate problems? At one point his frayed nerves snapped altogether. He stormed out of the caravan, and Sid later found him seeking comfort from the presence of his beloved snakes in the reptile house – still in the dark, because the electricity had not yet come on.

By the end of the day Sid was able to report for the six o'clock news that a number of animals had already been taken to the RSPCA Wildlife Unit in Somerset. Arrangements had been made to send thirty snakes to Chester Zoo, and homes had been found for various other small animals. He confirmed that the RSPCA would be meeting the cost of running the electricity supply until new locations had been found for all the animals.

Before Sid went home at the end of that long, trying day, he decided to carry out a systematic check on all the animals, birds and reptiles that were left in the zoo. He needed an accurate listing to see how many had still to be placed and which were most likely to have to be destroyed if no suitable alternative home could be found for them.

He went first to the enclosure which housed the three Himalayan bears. Although blind, Yogi, the thirty-five-year-old male, could still find his way around the enclosure. The casual visitor would probably not even have noticed his disability, but to move the old bear now from his familiar surroundings would cause him great distress, even if a home could be found for him and his companions, which Sid doubted. Following the recent closure of a bear park in Scotland, there had been a glut of Himalayan bears in British zoos which made the task of rehousing them all the more difficult, particularly one that was blind.

In the next enclosure Ricky the lion and Tank, the aptly named tiger, prowled backwards and forwards along the well-worn tracks they had made in their austere and bare cages. Next to them came the puma cage, where Zara lived. According to experts, she had reached the age of twenty and could be considered to be at the end of her natural life expectancy. Despite her blindness, she looked remarkable for her age, though the growth on the underside of her body should probably have been dealt with some time ago. Sadly, Sid placed a question mark against her name in his notebook.

He moved on to contemplate the noisy occupants of the adjacent enclosure. The family of dingoes that lived here – Bella, Cheyenne and Nuisance – kept up an almost constant howling. The fourth member of the dingo family, missing since Christmas Eve, had still not been found. The tree bears that occupied the tall tree opposite would not be difficult to house, thought Sid thankfully. Behind them was the sea-lion pool, in a very bad state of repair. Its occupant, a fifteen-year-old sea-lion with only one eye, named Nelson, had a further life expectancy of at least fifteen years. But where would he enjoy it?

Sid walked slowly on, stopping at various cages and enclosures, until he came to the fourteen-year-old lioness, Dandy-Leo. Her charming habit of rolling over on request had made her famous, so perhaps her antics would earn her a home. Sid had grave doubts about being able to place Roma, the ten-year-old Sumatran tigress, though, as there was a surplus of large cats in the country. The cards were stacked even higher against Kaffa, the old black leopard. He's lost most of his teeth, thought Sid – what hope is there for him?

Now to the monkey house. First Fluff, the green monkey, and then Grasshopper, the fourteen-year-old macaque that was in need of a lot of care and attention as he was recovering from hepatitis. Continuing around the monkey house, Sid next spotted Judy, a stumped tail macaque, aged about fifteen. A cheeky little creature, she'd steal your pen or pencil, given half a chance. Sid knew that, like a baby, she'd not settle for the night without her jumper to comfort her.

The bird house was next on the list. Rahja, the red macaw, greeted Sid with one or two well chosen words. I don't blame you, thought Sid. I feel much the same myself. He'd reached the end of his melancholy, self-appointed task. He gratefully accepted a cup of tea from Barbara Nyoka, then got into his van to drive the fourteen miles home. It had been an unusually demanding day.

Sue Jenkins, unaware of the drama her husband had been playing out, but well used to his long absences from home, had prepared a good hot meal for his return. As Sid walked through the front door he heard the headlines of the nine o'clock news. The demise of Knaresborough Zoo was being described to millions of people throughout the country. Within seconds, the telephone began ringing with one call after another from viewers who'd seen the item and wanted to offer homes to the animals.

Sue's meal never got eaten. The calls continued throughout the evening and into the small hours. Apart from the offers of help which had to be acknowledged and recorded, Sid was besieged by calls from the national newspapers wanting the full story. There's nothing like an animal item to bring the press to your door, as Sid was quickly finding out.

The following morning was Saturday. After a sleepless night Sid set off for the zoo again. On his way he was diverted by a radio message to Harrogate Police Station where he was told that the escaped dingo had been shot by a farmer, who had caught it worrying his sheep and mistaken it for a dog.

At the zoo he was met by David Hornsey. Using Barbara Nyoka's caravan, complete with telephone, as a base to work, the two of them set out to sort through all the offers of help that were still pouring in. They had to sift through long lists of names and telephone numbers before they could promise new homes for the zoo animals. Some of the offers could be accepted immediately, but others could not. David Hornsey spent his time checking with local authorities whether or not the callers had licences to cover them for keeping animals listed under the Dangerous Wild Animals Act. Not only zoos had telephoned with offers, but private collectors as well. Obtaining a licence to keep dangerous animals is, rightly, very difficult. Indeed, the two men would not have been sitting in Mrs Nyoka's caravan had Knaresborough Zoo been able to keep up with the stringent standards the Act requires to be met.

The offer of a special vehicle from Twycross Zoo to transport the animals to their new homes was welcomed by Sid Jenkins. At last Nick Nyoka appeared to have agreed to Galan the ape going to Dudley Zoo. At the same time, a wildlife park in Northamptonshire had offered to take three of the monkeys – Fluff, Grasshopper and Judy – and discussions were going on with a zoo in North Yorkshire that might be prepared to take the other two. Sid could hardly believe how well the day had started. Nearly all the primates had firm homes to go to, and it was now just a matter of arranging to get them there.

As the day progressed, further offers of help came in and Sid was beginning to feel that all was not lost. However, the bubble of optimism that was beginning to grow burst quite forcibly when he received a call from a debt-collecting agency, acting on behalf of a tourist magazine that had not been paid for an advertisement featuring the zoo. According to them, an earlier approach to Nick Nyoka for payment had met with the response that they should take the mandrill ape, and they now informed Sid that a restraining order had been placed on Galan, enabling them to seize him in lieu of the debt. Sid learned the warrant had been issued by the Harrogate County Court.

Late on Saturday Sid went home to spend most of the evening writing up the week's events for his file. On the following morning he went back to the zoo. It was Sunday, and much quieter, especially without the hordes of pressmen around. He had to check that all the animals were being fed: after all, the RSPCA was paying for the food and it was up to him to make sure it was being given to the animals. While there, Sid spent some time on the telephone following up the offers of homes that were still pouring in. One of the calls was from Hertfordshire where a leisure park with a licence to keep wild animals, wanted to offer a home to Dandy-Leo, the lioness. This was very good news; Sid had thought the big cats would be the most difficult of the animals to rehouse. After consulting with David Hornsey, arrangements were made for the lioness to be collected from Knaresborough on the following Sunday.

Monday morning came at last – another day of heavy snow. Sid was back at the zoo by 7.30 am, to find a message had already arrived from the County Court bailiff confirming the restraining order which had been placed on Galan. The ape could not be moved from the zoo without the court's permission. Sid and David Hornsey immediately taxed Nick Nyoka with what had happened. He laughed. Apparently, he had told the debt-collecting agency they could take Galan as a joke, never dreaming they would take him seriously. The only way to sort out the matter, Sid and David concluded, was to go in person to the County Court in Harrogate and attempt to get the warrant removed.

Slipping unnoticed in David's red Granada past the dozens of newsmen already stationed at the zoo gates, they encountered Paula, Barbara Nyoka's eldest daughter, on her way to her mother's caravan. It was her birthday and David wound down the window to wish her many happy returns of the day. The circumstances could hardly have been less auspicious, both men thought privately to themselves.

It took only ten minutes to drive the four miles to Harrogate. The bailiff's office was situated in a large open-plan office on the second floor of the modern court building. David Hornsey spoke first to the Chief Clerk. He told them that arrangements had already been made to move the mandrill ape to Dudley Zoo: a van was at that moment on its way to collect him. Not only that, two veterinary surgeons were due to arrive to sedate the ape, and the police had been informed because guns would be used to fire a tranquilliser into Galan. It was essential, in his opinion, acting on behalf of Harrogate Council, and in Sid's, representing the RSPCA, that immediate steps were taken to move the animal to ensure its continued safety. Any delay might mean his chance of a future home at Dudley Zoo would disappear – they had already had to move a mandrill ape to Southport Zoo to make room for Galan, and would not be able to keep the offer open indefinitely.

'Well, as far as we are concerned, you *can* move it. We don't want to know,' replied the County Court clerk. 'Mind you, it would be worth selling to raise the amount we need on the warrant.'

So far as that was concerned, David Hornsey told him, the ape was worthless. Sid Jenkins agreed. The animal might appear valuable on paper, he said, but if the court tried to sell it, they wouldn't find any takers. He pointed out that he and David had contacted every possible zoo and private collection in the country, and the only one that could do something to save Galan was Dudley. The only way that money could be raised from Galan was if he were killed, and his skin sold. That would cause a public outcry. This argument appeared to clinch the matter. Within a few minutes the County Court officer had agreed to withdraw the warrant.

Back at the zoo again, Sid's first priority was to make sure that Galan's removal to Dudley Zoo went forward without any further hitches. He got on the phone to the vets who were going to sedate him for the journey to tell them what the ape's weight was – they'd need to know this to work out the correct dosage to tranquillise him safely. He weighed 19 kilos.

Meanwhile the van that was to take him to his new home had arrived. It was driven by John Voss, whose expertise and knowledge of zoo animals was to be an enormous asset over the coming days in making sure they were transferred successfully from Knaresborough to their new homes. By now all the waiting reporters and camera crews had got wind that something was about to happen. Nick had already caused consternation among them that morning when he had stormed out of a discussion with Sid and David in one of his irrational bursts of temper to call a hasty press conference. All the remaining animals, he had announced, including the big cats and the other large animals, were going to be put down.

This, the assembled press did not want to witness, if and when it happened. However, to see the ape hit by a dart from a tranquilliser gun was a different matter, which would make good pictures for the front pages of their newspapers or find a high-ranking spot on the evening's television news programmes. As the two vets arrived, carrying their dart gun and a cardboard box containing the tranquilliser, the cameramen were fighting for vantage points around the enclosure, pushing at each other in their efforts to gain the best position.

Disturbed by the noise and bustle, Galan retreated to a dark spot behind his cage at the back of the enclosure. The two vets took advantage of this move to deprive the press contingent of their chance to capture the perfect picture. Although it was bright sunshine outside, it was dark at the back of the enclosure as one of the vets aimed the dart gun at Galan, out of sight of the waiting cameras. The ape was only four feet away as the vet steadied the gun with both hands and fired. The tranquilliser took swift effect, and Galan drifted peacefully off to sleep. It took the two vets, the driver of the van and the zoo assistant to carry the heavy animal to the waiting vehicle.

After Galan had been driven away, the press gathered round Sid for an update of the situation. One of the reporters asked what was going to happen to the other monkeys. Sid told him that at one time he had thought it probable that they would all have to be put down because they were old and were used to living on their own. Most zoos now prefer to keep monkeys in breeding colonies, but these would be too accustomed to being on their own to adjust easily to such conditions. However, a wildlife park in Northamptonshire had just offered homes for three of the five monkeys.

'Does that mean the last two will have to be put down?' someone asked. Sid just nodded his head. Efforts were still being made to find homes for the big cats, he went on. But most other private zoos had plenty of these animals already. In fact, overstocking was so great that some zoos had even started giving female cats contraceptive pills, or had vasectomised the males. In the end, it looked as if these animals, too, would have to be put down.

One of the television reporters asked Sid how he felt about this. He replied, 'Very, very sad. They are all marvellous animals; I know them well. I can talk to them and go in the cages with them. I'm sick. I joined the Society to help animals. It is going to be a sad day for me when we eventually have to put these large magnificent animals down. All this could have been avoided – they should never be in captivity.'

Like those that had preceded it, however, it was to be a day of quickly shifting moods. When Sid finally left the zoo that Monday evening his spirits were a good deal higher than when he had delivered those closing words a few hours earlier. Shortly after the interview had ended, an offer had come in for the two remaining monkeys. They were soon to take up residence in Flamingoland, a zoo in North Yorkshire. The apparently impossible had been achieved. Galan and the monkeys had all found homes. Perhaps the big cats could be relocated, too?

Even now, Sid could not relax and forget Knaresborough Zoo. The story of its closure was continuing to make the headlines on all the television news programmes, producing hundreds of telephone enquiries, not only to the zoo and to Harrogate Council, but to the RSPCA headquarters in Horsham, as well as to the area control room in Leeds. After office hours, all urgent telephone calls into the Group Communication Centre are re-routed to the duty inspector. Usually this is shared on a rota-basis among all the inspectors, but because of the emergency at Knaresborough, Sid was taking all the calls at home every evening. Often they went on well into the early hours.

Not all the calls were genuine. On Saturday night it had looked as if a home had been found for the lioness, Dandy-Leo, after a man had rung from the Great Yarmouth area, explaining that he had circus connections and possessed a large area of land on which he kept former circus animals. All such offers were first checked out by David Hornsey and his colleagues in the Harrogate Council office, as well as by the Wildlife Department at RSPCA headquarters, for any information they might have concerning the proposed new owner – whether a zoo or an individual. This caller turned out to be working as a chef on an oil rig. His sole animal was a llama which he kept tethered out on land in Great Yarmouth, much to the concern of the local residents. Now, on Monday evening, Sid received another call offering a home to Dandy-Leo. This time it came from Scotland from a man who claimed to own a private park in Edinburgh that could house the lioness. This call, too, was checked by the Wildlife Department and found to be a hoax. Sid and David quickly learned from these experiences never to take an offer at face value, never to build their hopes too high, but such senseless tricks, none the less, left a bitter taste.

But there was another side to the coin, as Sid was to find out later that week, when a young nine-year-old girl from Guiseley, a suburb of Leeds not far from Knaresborough, sat down in the bedroom of her terraced cottage and wrote a letter to Buckingham Palace. She had seen and heard many of the local and national news bulletins about the zoo, and had been so concerned about the plight of the animals that she had decided to ask the Duke of Edinburgh to try and do something to save them. She wrote:

> Dear Prince Philip
> I am very sad and upset because I heard on the news that Knaresborough Zoo is closing down. Some animals will have to be put down because they are old and nobody wants them. I know you care about animals and nature, that's why I am writing to you. Please could you do something to save their lives?
> Love Nadine – aged 9

She posted the letter and waited.

Meanwhile, at the zoo, arrangements had been made

with suppliers to provide meat for the big cats and other carniverous animals, and fruit and vegetables for the bears. This food was being paid for by the RSPCA. It was agreed that Friday should be the deadline to decide the future of the animals, and arrangements were put in hand to draw up legal documents to arrange for the disposal of any animal that had not been given a new home by that date. A veterinary surgeon specialising in zoo animals was approached, and agreed to make himself available when the time came to put the animals to sleep.

All this was done, as so much had been that week, with at best only half-hearted assistance from Nick Nyoka. He still seemed unable to face up to the reality of the situation and at this stage was pinning all his hopes on a Yorkshire businessman whom he believed was ready to take over the debts of the zoo and keep Nick in business. Sid Jenkins had good reason to doubt the truth of this story, and he and David pressed on with their efforts to rehouse the remaining larger animals.

Things were beginning to look more promising on this front. Martin Grantham, a computer expert, now approached Sid with a proposal to save Roma the tigress. When he had first heard about the fate of this lovely creature on television, he had felt compelled to try and do something to save her, and had got in touch with Len and Diana Simmons, the owners of Linton Zoological Gardens near Cambridge. What was the possibility of their being able to house the tigress at Linton? Len Simmons replied that although they had plenty of ground space, they did not have an enclosure sufficiently large and built to the required safety levels to keep Roma.

Martin was not easily put off. What if the money could be raised to build an enclosure and sleeping quarters? Would Len be prepared to have the animal then? He would. The two men immediately began to work around the clock to raise the money for their project. Help came from an unusual source – the RAF. Sergeant Barney Barnes is a member of the newly-formed 74 Squadron RAF, based at nearby Stowmarket in Suffolk. The Squadron's nickname is the Tigers. When he heard of Roma's plight, Barney persuaded his fellow airmen that they should adopt her as the Squadron's mascot – they did so, and offered to pay for Roma's keep at Linton Zoo for the rest of her life. Extra donations also poured in to BBC Radio Cambridge after they broadcast an appeal to save the tigress. In fact, so much money was raised, that not only were Martin and Len able to bring Roma to Linton, they could provide a home there for Kaffa the black leopard as well.

The future of these lovely animals ensured, Sid now launched an appeal to save Satan, the old lion, whose enclosure was near the public entrance to the zoo. Suffolk Wildlife and Country Park had already offered a home to the four dingoes and Sid asked the director if he would be prepared to have an enclosure built to house Satan as well, if enough money could be raised. This he agreed to do, and an appeal fund was set up at the National Westminster Bank in Knaresborough. Enough money was secured to allow Satan, some time later, to leave for his new home in the south of England. This much had been achieved by Thursday, the day before the agreed deadline ran out.

Late that morning Sid and David were meeting with Nick Nyoka to try and convince him once and for all that there was going to be no last-minute rescue by a local businessman. Halfway through their discussions the telephone rang. 'It's Buckingham Palace, for you, Sid, it's the Duke of Edinburgh's office,' said someone.

'Do you mean the Duke of Edinburgh's Award Scheme office?' asked Sid, slightly puzzled, wondering why they would be ringing him at the zoo. 'It must be some course they'll be wanting me to organise,' he added. He does quite a bit of work with the Award Scheme.

He was wrong. It was the Duke of Edinburgh's private office at Buckingham Palace. Little Nadine's letter had been received and the call was being made on Prince Philip's instructions to ask Sid to contact Nadine on behalf of the Prince and explain the situation at the zoo to her. This call from the Palace was just the fillip Sid needed to help him to get through the depressing business that lay ahead. He decided to call at Nadine's home on his way to the zoo the next morning. He was able to tell her that nearly all the animals had been saved, and that everyone was still doing their best to find homes for all of them, and received an enchanting smile in return from the little girl.

Later that day more vehicles arrived at the zoo on what was now the almost routine task of transporting animals to a new home in another part of the country. This time it was the turn of the family of tree bears and Nelson, the one-eyed sea-lion, who were going to Scarborough Zoo and Marineland. The move went well until the last moment. One of the tree bears decided that he was not going to be moved from Knaresborough. He climbed the highest tree and perched up there, refusing to come down, despite every blandishment, for a couple of days until hunger finally persuaded him to give himself up. He was belatedly sent on his way to join his old companions in their new seaside enclosure.

That Friday evening Sid rang the specialist vet to confirm that he would be at the zoo first thing on Monday morning to put the last animals down. It was a sad decision, but one that Sid knew could not be put off any longer. Funds would no longer be available to provide food and heating for the animals that remained unhoused. He spoke to the RSPCA headquarters to arrange for two members of their Wildlife Department to be present to assist him in this final melancholy task.

Sunday, 12th January. Another very cold day. As David Hornsey arrived at the zoo well wrapped up in a large overcoat and sporting a deerstalker, Sid greeted him almost cheerfully. 'Good morning, David, I've some good news. As of yesterday, twenty-five out of the twenty-eight animals have been found homes. What we have left now is the puma and the two adult bears.' At this stage it looked as if both the remaining lions, Dandy-Leo and Ricky, had been found homes.

'I've had a telephone call from someone who is prepared to save all the bears,' David Hornsey replied. 'He is looking at the possibility of rehousing all three bears.' It was good news. The men went into the caravan to ring back and find out more. But when David Hornsey put the phone down he was looking less happy about things.

'Well, Sid,' he said, 'I don't honestly think it's on. He wants to house the bears on his own land and he hasn't got a licence for wild animals. He's some sort of market gardener and uses a bear as his logo – he'd like to use the bears for advertising. I've been talking to his agent and asked him to get his client to give the situation a realistic appraisal, and call us back with an answer within the next hour.'

Sid sighed. 'It doesn't sound too hopeful.'

'It's obviously an offer made from the bottom of his heart,' David said, 'but he doesn't understand the implications of what is needed to keep a wild animal properly.'

'At least we are still trying,' replied Sid as he put on his peaked cap to leave the caravan.

By early afternoon, Tim Thomas and Paul Vodden had

arrived from the RSPCA Wildlife Department to lend a helping hand. One reporter, Bruce Smith from the *Yorkshire Evening Post*, was there. He had been at the zoo every day since the story first broke. From time to time during the previous fortnight he had put down his notebook, rolled up his sleeves, and helped with the removal of the animals. Today was going to be no exception. By now he had built up a great friendship with Sid who trusted him not to crash in and out of the zoo looking for sensational angles, like some of his colleagues.

Next to arrive was John Voss with his large white van to collect Dandy-Leo the lioness for her new home in Hertfordshire. The whole afternoon was to be spent in coaxing Dandy into the special transportation cage which John had prepared. The metal cage was six feet long and about three feet high, but only two feet wide, so that the large animal would not be thrown about in transit. It was fitted with two runners similar to those on a sledge to help manoeuvre it, but even so, it was so heavy that it took six men to move it into position up against the door of Dandy's inner enclosure. Once there, the end of the cage was lifted up to allow room for the lioness to crawl inside. But first she had to be tempted in. One of the zoo assistants hung a large piece of red meat at the end of the cage, opposite the open gap. The big metal door to Dandy's enclosure was pulled back and the group standing around waited to see if she would enter.

Nothing happened for about ten minutes, then the lioness popped her head out of her enclosure, held her head in the air and had a sniff. Twice she got halfway into the cage to investigate the smell of meat, then backed out into the safety of her own enclosure. Sid crouched down next to the cage and started calling her. Nearly all animals respond to something in the tone of his voice. 'Dandy . . .

come on Dandy . . . isn't this nice. Come on . . . that's a nice girl.' Dandy growled, came out, and took the meat. The cage door was pulled down shut behind her. She was in.

At first she was very agitated. Unlike Galan, she had not been tranquillised. Because of her age, it was felt that would be taking too big a risk. The plan was to leave her for an hour to settle down and become accustomed to her new cage before loading her into the back of the van. It was dark when the time came to move the cage. Together, the lioness and the large metal cage must have weighed more than half a ton. All available hands, including Bruce Smith from the *Yorkshire Evening Post*, helped to drag and push the cage towards the waiting van. It took nearly half an hour to get it the twenty feet from the enclosure and into the back of John Voss's vehicle. Once inside, Barbara Nyoka knelt over the cage to say goodbye. It was heartbreaking for her. She had known the lioness for fourteen years.

Whatever satisfaction had been gathered from seeing the old lioness on her way to a new home was soon dispelled by the news that, after giving the idea further thought, the market gardener had decided that he could not provide a home for the bears after all. It had been their last chance. Now nothing could be done.

Sid Jenkins did not sleep at all that night. He had been dreading what the next day would bring since the moment he had heard that Nick Nyoka had lost his appeal.

He was at the zoo by 6.30 the next morning, wanting to get there before the entourage of journalists and television crews arrived. Although they had given assurances earlier that they did not wish to record the animals being put down, Sid felt they might go back on their word, determined to secure sensational pictures for their papers. The police were also there to ensure that the press contingent remained outside the zoo and to be on hand in case of demonstrations by animal extremist groups.

Just before 8 am, a large yellow vehicle belonging to Harrogate Borough Council reversed into the zoo carrying two waste disposal skips. A man in a red fluorescent jacket waved the vehicle up the roadway, past Barbara Nyoka's caravan, into a position between the bears' cage and the enclosure containing Zara the puma. Sid just stood and watched with his hands in his pockets. At that moment the white van that had taken Dandy-Leo away the night before turned into the zoo. The lioness was still inside. Apparently her new owners had had a change of heart. So Dandy-Leo was back again at Knaresborough. Nick Nyoka arrived and peered through the rear window of the van and shouted to Dandy that she was back home again. He was the only one who looked pleased. A couple of days later Dandy-Leo was given a home at Flamingoland Zoo in North Yorkshire.

But now Nick's smiles turned to tears. He had been housing a baby puma for some weeks in his caravan. It had been ill and had not been visited by a vet. That morning he had found it dead. Sid knew about the puma but was not aware of its condition. Nick and Sid walked back together to the caravan. A few minutes later Sid emerged looking very angry, carrying in his arms the body of a dead baby puma. A post-mortem later revealed that the cub had died from enteritis.

A JCB excavator arrived and parked next to the Council skips. Finally a white van turned into the zoo's gates. It was driven by the vet and contained the equipment that would put four of the animals to sleep – Zara the puma, the two adult bears, Yogi and Dolly, and Ricky the lion. Treasure, the youngest of the bears, had been saved: she was going to a zoo in Scotland and would be collected later. The previous evening the owners of the zoo who had promised Ricky a new home had rung to cancel the agreement. They had been threatened by action from the Animal Rights movement if they took him in and, reluctantly, had been forced to go back on their word.

The vet walked to the enclosures to see the animals and assess the correct lethal dosage that would be needed to put each down. His tall figure, wearing a light jacket over a suit, sported slightly incongruous-looking yellow yachting boots. Now he got to work with the array of specialised instruments that were laid out in the back of his van. First he put the drug into a syringe and then placed the syringe in a spring-loaded dart. The whole unit then went into the end of a blowpipe.

Just at that moment Sid came running up, shouting, 'Hold it, hold it.' He'd just received a telephone call from Belfast Zoo, saying they might be able to offer Ricky a home. The vet continued to prepare his instruments. There was no hope of saving the puma and the two bears. Sid went back to the caravan. Within minutes he emerged and grimly informed the vet that the call from Belfast had turned out to be a false alarm.

Sid watched pensively as the vet walked over to the first animal, Zara the puma. Virtually blind, but just able to make out shapes, she was lying on top of her cage. The vet slid the blowpipe through the wire mesh of her enclosure. He took a deep breath and blew. The dart hit Zara just behind her left foreleg. She jumped down into the enclosure with the dart still in her. Sid turned away.

Ricky the lion was next. The vet fired the dart into his left hindleg. He slumped near the fence of his enclosure, enabling the vet to inject him with a further dose. Sid stood and watched. 'It's sad, isn't it,' he said, simply. The vet nodded in agreement.

All this time Nick Nyoka remained in his caravan. The Council workmen who had driven the vehicles into the zoo walked up and down with the two policemen, their hands behind their backs. The bears' enclosure was between the puma's and the lion's cages. They had sensed trouble and were pacing agitatedly up and down their enclosure. On the other side of the perimeter fence the television and press crews were becoming impatient at being kept locked out. Some of them had climbed up nearby trees and had been pursued by police officers to bring them back to earth.

Inside the zoo one of the dingoes, still awaiting collection by its new owners, started to yelp. It had seen its neighbour Zara put to sleep and its piercing whine could be heard half a mile away. The sinister sound accompanied Paul Vodden and the Council workmen as they went into Zara's enclosure to carry her now lifeless body out. The three men lifted her by the legs and took her to one of the skips. Nick Nyoka appeared, carrying his Polaroid camera. As soon as the animal was in the skip he snapped a photograph. He was obviously dazed and shocked by what was happening.

Ricky's body was so heavy that it had to be carried on a stretcher to the skip. The vet meanwhile had put the two bears to sleep and now the JCB excavator started to knock down the concrete breeze-block wall that separated their enclosure from the public. The vet checked that the bears had gone down peacefully and gave the signal for the bodies to be removed. A rope was tied around Yogi's body and attached to the rear axle of the excavator. It dragged the bear from the enclosure out onto the zoo roadway. Because the eager eyes of the press were peering through the cracks in the main gate, those present inside the zoo formed a human wall to prevent the cameramen recording the unceremonious sight of the bear being dragged up the roadway to the skip. Once there, the JCB shovelled the bear inside. The same awful treatment awaited Dolly.

In another ten minutes it was all over. The skip contained the limp carcases of Zara, Ricky, Yogi and Dolly. They were loaded onto the back of the waiting lorry and a tarpaulin placed over them. Finally, Sid climbed up the rear of the skip and placed the body of the puma cub inside.

Nick Nyoka turned and walked away, alone, back to his caravan. Only a few animals remained now, awaiting collection by their new owners. They would all be moved in the coming days. Sid Jenkins stood for a moment at the side of his van. He took a long, last look at the zoo, staring especially at the empty cages that only a few minutes earlier had housed animals that had given joy to thousands of children and their parents for over a decade. There's nothing wrong with animals, he thought, just people. Then he stepped into his van and drove away. He had tears in his eyes.

# RSPCA COMMUNICATION CENTRES

BIRMINGHAM (021) 427 6111
Birmingham

BRIGHTON (0273) 685146
Billingshurst, Crawley, Horsham, East Grinstead, Hastings, Eastbourne and Brighton

BRISTOL (0272) 40640
Swindon, Devizes, Chippenham, Bath and Bristol

CAMBRIDGE (Newmarket) (0638) 742492
Peterborough, St Ives, Kings Lynn, Thetford, Bury St Edmunds and Cambridge

CANTERBURY (0227) 457733
Ashford, Folkestone, Dover, Ramsgate and Canterbury

CARMARTHEN (0267) 233954
Haverfordwest, Llanelli, Swansea and Carmarthen

CHELMSFORD (0245) 422202
Chingford, Harlow, Braintree, Colchester, Southend, Upminster and Chelmsford

CHELTENHAM (0242) 570510
Cinderford, Gloucester, Stroud and Cheltenham

CHESTER (0244) 374282
Birkenhead, Ellesmere Port, Crewe, Chester and Macclesfield

GREAT YARMOUTH (0493) 603264
Norwich, Ipswich, Lowestoft and East Dereham

GUILDFORD (0483) 505735
Kingston-upon-Thames, Woking, Epsom, Redhill and Guildford

KENDAL (0539) 29012
Carlisle, Whitehaven, Dalton and Kendal

LEAMINGTON SPA (0926) 312974
Nuneaton, Coventry, Rugby and Leamington

LEEDS (0532) 629838
Skipton, Harrogate, Halifax, Huddersfield, Bradford and Leeds

LEICESTER (0533) 877766
Loughborough and Leicester

LINCOLN (0522) 39977
Grimsby, Boston, Grantham, Lincoln and Scunthorpe

LIVERPOOL (051) 264 7496
Southport, St Helens, Warrington, Liverpool and Wigan

LLANDUDNO (0492) 78866
Anglesey, Bangor, Rhyl, Llandudno, Wrexham and Colwyn Bay

LONDON
South of the Thames (01) 228 1131
North of the Thames (01) 228 0656

MAIDSTONE (0622) 675855
Dartford, Maidstone, Sittingbourne, Gravesend, Sevenoaks, Isle of Sheppey, Tonbridge and Tunbridge Wells

MIDDLESBROUGH (0642) 224641
Darlington, Durham, West Hartlepool, Northallerton, Middlesbrough and Stockton-on-Tees

NEWCASTLE (Dunstan) (0632) 608899
Alnwick, Hexham, Sunderland and Newcastle

NEWTOWN (0686) 25075
Aberystwyth, Barmouth, Kington

NOTTINGHAM (0602) 784965
Nottingham and Newark

OXFORD (Wheatley) (08677) 3006
Banbury, Oxford, Aylesbury, High Wycombe and Milton Keynes

PERRANPORTH (087257) 3024
Cornwall

PORTSMOUTH (0705) 737058
Southampton, Fareham, Portsmouth, Petersfield and Chichester

PORTH (044361) 6975
Aberdare, Merthyr Tydfil, Port Talbot, Bridgend and Cardiff

PLYMOUTH (0752) 786532
Barnstable, Okehampton, Exeter, Torquay and Plymouth

PRESTON (0772) 22463
Lancaster, Blackpool, Fylde, Morcambe, Preston and Accrington

READING (0734) 509341
Newbury, Reading, Fleet, Basingstoke, Slough, Bracknell and Windsor

RIPLEY (0773) 42198
Chesterfield, Mansfield, Derby, Burton-on-Trent

SALFORD (Manchester) (061) 7361231
Bolton, Bury, Oldham, Hyde, Stockport, Altrincham and Manchester

SALISBURY (0722) 331872
Winchester, Andover, Lymington, Bournemouth and Salisbury

SHEFFIELD (0742) 753996
Sheffield, Barnsley, Wakefield, Doncaster and Rotherham

SHREWSBURY (0743) 245228
Oswestry, Bridgnorth, Stretton and Shrewsbury

STAFFORD (0785) 46642
Newcastle-under-Lyme, Wolverhampton, Walsall, Stafford and Stoke-on-Trent

STEVENAGE (0438) 62783
Kettering, Northampton, Luton, St Albans and Bedford

WORCESTER (0905) 26504
Kidderminster, Redditch, Worcester and Hereford

YEOVIL (0935) 29096
Cheddar, Taunton, Yeovil, Dorchester and Blandford

YORK (0904) 39991
Scarborough, Bridlington, Hull and York

If you have any difficulty in tracing your local Inspector, the name and telephone number can always be provided by the RSPCA Branch or Animal Home locally or by RSPCA Headquarters, The Causeway, Horsham, Sussex RH12 1HG. Telephone Horsham (0403) 64181

ANIMAL SQUAD

*Sid Jenkins* joined the RSPCA as a trainee inspector in 1973 and was appointed an inspector in April 1974, when he was posted to Wales. He then served as the Inspector for Harrogate before being promoted to the rank of Chief Inspector in September 1979. He moved to his present station in Leeds in April 1980. Whilst in the North he has written a weekly column on animal care in the children's section of the *Yorkshire Evening Post*, and has recently taken over the Monday night Pet's Corner slot on the same paper. He lives with his wife Sue on the outskirts of Leeds.

*Paul Berriff* started his career as a press photographer on the *Yorkshire Evening Post* before joining BBC Television News and the regional magazine programme *Look North* in 1967. Two years later he won the Documentary of the Year Award with his 'World About Us' film, *The Great Unknown*, about the first canoe expedition through the Grand Canyon. Now an independent producer, he specialises in outdoor and adventurous subjects and has made many major single documentaries, among them *Gold from the Deep*, the story of the gold bullion salvage operation from the sunken warship HMS *Edinburgh*. His series include *Rescue Flight*, *Medic-One-Six*, *Lifeboat* and *Lakeland Rock*. He has won nine major film awards. Paul lives with his wife and two children on Humberside and carries his action-packed life outside the world of films by serving with HM Coastguard as Auxiliary Coastguard-in-Charge of the River Humber.

# Acknowledgements

I would like to thank the following for the parts they played in my adventures in Nepal and the production of this book:

Claire Westley, whose counselling sessions inspired me to start writing in the first place. Mark Dunn, of AFL Financial Services Ltd, for keeping tight control of the purse strings. My editor, Caroline Swain, for her tireless efforts to turn me into some sort of a writer. Abhishek Shahi for his line drawings of family members (other illustrations are by me). Tonya Stewart, who worked on my first draft and gave me so much encouragement.

Above all, my partner Adrienne Wilson for her exemplary proof-reading but, more importantly, her forbearance when I was in danger of becoming obsessed with all things Nepalese.

# 'Discovered in Kathmandu'

In Kathmandu, a retired English schoolteacher is taken to an orphanage where he meets a young man who desperately wants to train as a doctor. After a year's agonizing indecision, the Englishman takes up the challenge and promises to sponsor him. This leads to a life-changing experience in which he himself comes to be 'adopted' into a Nepalese family as brother, uncle, friend and godfather.

This book offers a unique and warm-hearted look into the dilemmas of low-caste family life living on the brink in Nepal. Meet the penniless tailor, Kiran, who was saved from suicide; learn about the comic informality of house-hunting in Kathmandu; get to know a colourful array of characters, cheerful and courageous in the face of extreme poverty. The finale of the story is a trip to Pokhara, the trekking capital of Nepal, for 12 family members and the author himself.

# About the author

Nick Morrice had the benefits of a privileged upbringing and private education in the south-east, followed by three years at Durham University. A teaching career followed, culminating in a twelve year appointment as Head of English at Downsend Prep School in Leatherhead. Following early retirement in his mid-50s, Nick became interested in charity work, and founded the West Surrey Local Support Group for WaterAid. He also sponsored a Nepalese boy through a local charity, 'CHANCE for Nepal', which led him to venture into the country's capital city, Kathmandu, for the first time in October 2009.

As it happens, he had inherited some money following the death of his mother in the spring of 2008.

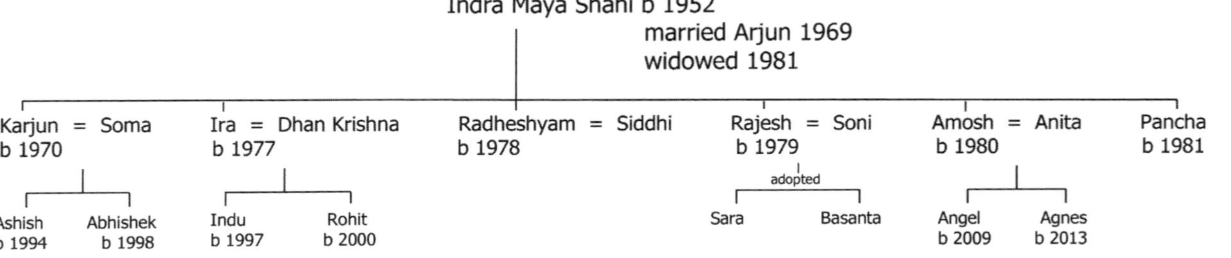


**Orphans in Wasta Care Centre**

| | | |
|---|---|---|
| Niran Karki | b 1989 | Studying Chartered Accountancy in Delhi |
| Ramesh Khadka | b 1990 | Studying Medicine at Lumbini Medical college |
| Anish Shahi | b 1992 | Studying Hotel Management in Kathmandu |
| Sunil Karki | b 1992 | Head Boy of Oxbridge College, Kathmandu |
| Arun Shahi | b 1996 | Student at Oxbridge College, Kathmandu |

# Foreword

This is the story of five young men who have been brought up together from an early age in an orphanage in Kathmandu. Strictly speaking they are not all orphans: Ramesh has a mother living in a distant rural area, while Arun and Anish have a father living in Kathmandu. But their parents either could not afford to support them, or simply abandoned them when they were young.

I refer to them as "boys" throughout because that is how they refer to themselves, and how their adoptive family refers to them. At the date of publication they are aged between eighteen and twenty-five, but I suspect they may be referred to as "the boys" for a long time yet, at least until they are married. I only mention this in case it would seem I am belittling them. Nothing could be further from the truth. Likewise, I am not their uncle, although they call me that. But I am their unofficial godfather – that was their idea, not mine – and I am quite happy with it.

Pokhara
Ghorahi
Tansen
Butwal
Lumbini
Kathmandu
Patan (Lalitpur)
Nagarkot
Bhaktapur
Birganj
INDIA

# Dedication

*This book is dedicated to Barbara Datson,*
*who opened my eyes to the wonders of Nepal.*

# Chapter One

It is night. I am being driven along narrow lanes at a furious pace by the hotel chauffeur. He honks his horn constantly while conducting a one-way conversation with me, because after my long flight I am too tired to reply. The road turns into a bumpy, unmade lane which does not bode well for my lodgings. But the hotel turns out to be a welcome haven of calm, genteel order and I am greeted cordially by well-groomed sentries at the wrought-iron entrance gates.

"Welcome to the International Guest House." The young man at the desk hands me the key to my room and a porter is instructed to take my bags upstairs.

I have arrived in Kathmandu, a place so remote from my home in South-East England – and so associated in my mind with the exotic and mysterious world of the East – that it feels slightly unreal to be here at all. But it is of course totally real, and after taking a shower, I embark on my first venture into its noisy, dusty streets.

My hotel is situated on the edge of Thamel, which I soon learn is the tourist area. It is a maze of confusing streets, brightly-lit shops selling pashminas, jewellery and paintings, and crowds of shoppers and shopkeepers vying with each other for the best possible prices. It is a seductive but confusing area and I am soon lost in its twisting warren of lanes. The air is sultry, and the combination of heat and dust makes my walk through the crowds a sweaty, uncomfortable experience. Being early October and therefore festival time, fireworks add to the general cacophony of beeping cars and motorbikes, rickshaws and cyclists, as well as loud piped music. I am glad to find my way back to the haven of my hotel.

The next morning I am able to appreciate its ornate splendours such as the heavy wood carvings, so typical of traditional Nepalese art, and plasterwork ceilings; outside, the grassy courtyard is adorned by a central brick stupa surrounded by elephants made of terracotta. Purple bougainvillea trails down from the upstairs landing, while the petunias in tubs remind me of home. Many of the staff are wearing marigold garlands in celebration of Dashain, the name of the current ten-day festival. I enjoy a breakfast of mango juice, tea and toast in another courtyard which surrounds an ornamental pond and, sitting in the warm sunshine, I collect my thoughts about the coming day.

Primarily, I have come here to do some teaching. I sponsor a 10 year old boy called Manish, who is being educated at Triple Gem School, and I have decided to meet him as well as do some teaching at his school. But I also want to see Kathmandu, so I have arranged for a guide to show me round the city. His name is Kaji Shrestha and I know of him because he was the guide of a trekking party from Canada in which my sister-in-law's piano pupil, Alice, took part.

His story is an unfortunate one. He is of Sherpa stock and had been a walking guide all his working life. But on one trip, when he was in his forties, he fell sick at high altitude and had to be airlifted back to Pokhara where the walk had started. The exact cause of this sudden sickness remained a mystery, but it was serious enough to ground him and put him out of work. Now this, for a man in the prime of life and with a wife and five daughters to support, was nothing less than disastrous, since there are no disability benefits or pension arrangements in Nepal to help him. His wife was earning a small living running a shop at the front of their house, but that was all they had.

I had heard about Kaji just before leaving England, and I considered that if I could give him some work guiding me round Kathmandu, this would benefit us both. It gave me great peace of mind to know that I would not be on my own in this mysterious city, and Kaji would probably be grateful for some income. He is due to meet me this morning, as is the headmaster of Triple Gem School, Lama Kondan. So I feel relaxed as I sit waiting for them in the hotel lobby.

I don't have long to wait. At 8:15, Kaji and his second daughter, Amita, arrive and the hotel receptionist introduces them to me.

"Hello. Are you Nick?" asks Amita shyly. She is a striking girl in her early twenties and is wearing a red sari and a few touches of make-up. Her father is short and stocky, well-built for a lifetime of guiding western trekkers in the Himalayan foothills. He speaks quite good English but needs Amita for our initial introductions.

Amita tells me she is a student of English language and literature at the Active Academy College and is hoping to become a journalist. She asks me about myself and I explain that I have been a teacher of English for most of my career. She tells me about the literature she has studied, including the plays of Shakespeare and I tell her about the Globe Theatre which I have recently visited, even drawing a diagram of it for her in my diary by way of explanation. I ask her if she has a computer – she doesn't, which puts me in mind of being of some practical help to Kaji's family one day. Kaji sits silently nearby.

This somewhat subdued start is enlivened at 9 o'clock by the arrival of Lama Kondan.

"Nick! Nick! Welcome to Kathmandu! It is so lovely to see you after all our emails. Are you all right after your flight? Are they looking after you well here?" and with a great flourish he produces the customary white silk scarf which he places round my neck. We bow to each other, palms together in the traditional manner, and I feel truly welcomed.

"Please introduce me to your friends," which I do and they talk to each other for a while in Nepalese. Lama Kondan is a Buddhist monk, but his manner reminds me more of the Chaucerian version: moon-faced, barrel-bodied and with a beaming smile. He cuts an imposing figure in his purple robes, and I can imagine him making a formidable and much respected headmaster.

After a few minutes he turns back to me.

"I have some unfortunate news. Your sponsored boy at Triple Gem, Manish, has not made it back to school for this term. There has been an incident in his village in the Terai region, a murder I believe, and as is usual in these situations, the police have roadblocked the area. I am so sorry about this."

I feel a little crestfallen but not totally downcast. At least I am here in Kathmandu and have met up with these people as arranged. I can't expect everything to work out exactly to plan. We talk about when I will start teaching, and I agree to come on Thursday and Friday for a few hours each morning. Lama Kondan and Kaji discuss the location of the school, and Kaji's first job will be to show me how to get there on foot.

"So!" Lama turns to me with a beaming smile. "I am glad you are here, Nick – you will be in safe hands with Kaji for the next two days. We will look forward to seeing you at Triple Gem on Thursday morning", and with a sweeping bow he makes his departure.

I am delighted to have two days' acclimatisation before my teaching begins and am looking forward to my first guided walk. Amita now bids farewell as she must go to college, but hopes to see me at her house in a few days when I will meet her mother and sisters.

Kaji and I now set off on foot in search of Triple Gem School, which I gather is about twenty-five minutes' walk away near the great Buddhist temple of Swayambunath. The day is warm, the streets are densely crowded and the sights before my eyes are endlessly fascinating in this very public city. Taking in all its sights and sounds will become my favourite pastime: the roadside shopkeepers selling their wares on the pavements or in open doorways; the bicycles, motorbikes and rickshaws dodging their way past the teeming pedestrians; men playing chess or ludo on the roadside, totally oblivious to passing traffic; the washing of clothes in the public laundry, the collecting of water, the washing of hair under a tap. It is all so different from anything I have experienced before. It makes my home town in England seem a trifle dull by comparison.

We find Triple Gem School quite easily and, with the help of a map, realise that there are several ways for me to walk here each day so I won't get lost. The Swayambunath Temple is a useful landmark and Kaji takes me to visit it now. Legend relates that the Kathmandu Valley was once a lake and that the hill on which the temple stands was 'self-arisen' *(swayambhu)* like a lotus leaf from the muddy waters of the lake. From its hilltop setting, Swayambunath offers fine views over the city and surrounding countryside. I admire its highly ornamented golden doorways, the vast Buddhist prayer wheels and the exquisite wooden carvings. The steps up to the temple are lined with craftsmen and the trees are crowded with chattering monkeys, giving rise to its popular name of the Monkey Temple.

Kaji then organises a taxi ride to take us to Bodhnath, home of one of the world's largest Buddhist stupas. This is the religious home for Nepal's large population of Tibetan exiles, and the streets throng with maroon-robed Tibetan monks. The vast whitewashed dome of the stupa is spectacular, as are the colourful paintings depicting the strange beasts and gods of the local mythology. A constant chanting of *Om Mani Padme Hum* fills the air as we walk around – and looking up I see a balcony where young monks are craning their heads to behold a sight which captures their gaze below. Dancing girls! Four exotically dressed young ladies are twisting their bodies, hands and feet to the accompaniment of a vocal and percussion band of four young men. Well, why shouldn't the novice monks be as entranced as I am?

There is a wide pathway surrounding the temple, which one is expected to walk in a clockwise direction. On the side opposite the stupa, there are several restaurants, and Kaji takes me to the upstairs terrace of one nearby which has a good view of the temple. Prayer flags stream out in colourful, fluttering profusion in all directions from its pointed top, while around the base of its circular mound are 108 small images of the Buddha (according to my guide book, 108 is an auspicious number in Tibetan culture). There is not much agreement on how old the Bodhnath site is, but it seems likely that the first stupa was built some time after 600 AD when the Tibetan king, Songsten Gampo, had been converted to Buddhism by his wives. Legend relates that he constructed it as a penance for unwittingly killing his father.

Kaji orders us a lunch of cheese and tomato panini with salad, accompanied by iced tea with lime. While we enjoy our meal I absorb the atmosphere of this sacred place, so colourful with its long lines of prayer flags waving in the breeze.

From here we head to Pashupatinath, the most important Hindu temple in Nepal. It is located on the banks of the Bagmati river, and my attention is drawn to the burning *ghats* used for very public cremations, which are accompanied by the equally public wailings of the bereaved. Acrid smoke fills the air. We in the West expect our cremations to take place out of view and public mourning to be restrained, but the tradition here is quite different. People give full expression to their feelings without any restraint or embarrassment. Death is simply a natural part of life.

I feel I have seen quite enough for my first day. Back in Thamel, Kaji bargains for some fresh fruit for me – and we part company, both delighted that everything is working out so well.

"I shall look forward to seeing you tomorrow, Kaji," and I give him a £20 note, the agreed fee for a day's tour guiding. I feel it's a bargain, and I suspect he feels equally pleased.

In the evening, I have one more planned meeting. This is with Pastor Shyam Nepali and his son, Amit. Shyam is pastor of the Nepalese Christian Church called simply Golgotha, and I had been given his name and email address by the Anglican Communion Office in London. I felt it might be an interesting contact, little suspecting what a momentous change it was to make to my life.

They arrive at about 6:00 pm by motorbike. We greet each other and shake hands as they stumble through a little English and I begin to wonder where this can possibly lead. I suggest that we might go out for a drink. We saunter out of the hotel precincts into the noisy and now dark local lanes. They are clearly as lost as I am, since they come from Patan – a town a mile or two south of Kathmandu – so maybe it is best to return to the comfort of my hotel. I order three Cokes which duly arrive but before we touch them, the pastor offers up a prayer:

"Heavenly Father, thank you for bringing us safely together on this joyful day. We thank you also for the refreshment of this drink which we share together. We offer it in Your service, in the name of Jesus Christ. Amen."

Now we can enjoy our drink with a clear conscience. The main topic of our conversation is how Shyam will pick me up on his motorbike at 8:00 am on Saturday (their worship day) and drive me to Golgotha Church.

"And we would like you to preach to us, please, in the way you minister to your church in England."

I gulp slightly at this, not being a church minister and therefore not having a handy sermon up my sleeve. But I think it best not to reveal this and politely accept the invitation. I promise to have my address ready for tomorrow and will hand it over to Amit at a pre-arranged meeting place. This will give Amit time to prepare a translation for the congregation when I arrive on Saturday.

I calmly bid them farewell, then return to the hotel in the hope of divine inspiration. After taking a refreshing shower, I sit in the tranquil coolness of my room. The dark ceiling beams and wall panels lend it a restful atmosphere and I reflect quietly on the events of the day. I am prepared for teaching young Nepalese schoolchildren, but not for preaching to a Nepalese church community. But this will probably be a one-off occasion and I confidently believe I can rise to the challenge. It is, however, to have unexpected consequences which will go well beyond the original intentions of my trip.

✸ ✸ ✸ ✸ ✸

# Chapter Two

Feeling debonair and light-hearted the following morning, I branch out into pancakes with honey for breakfast, then go and buy a peaked sunhat inscribed "Yak Yak". The weather is altogether warmer than I had anticipated. It must be in the mid-80's, but it is a dry heat so not uncomfortably hot. I am already getting a sense of how polluted Kathmandu is, and my throat is beginning to rasp from the traffic fumes. I notice a number of pedestrians wearing mouth masks to protect themselves from the worst effects. Kaji is in the street outside the hotel at 10 am as arranged, and I immediately tell him the initial plan for the day: we will take a taxi to meet Amit half-way at the Prayag Pokhari church. Kaji finds a taxi for us, but the plan is not so easily fulfilled as neither we nor the taxi driver know where the church is. We drive along some impossibly pot-holed roads and eventually a map is consulted. After a lot of guesswork and more impassable 'roads', we eventually arrive and have only to wait a few minutes before Amit arrives. He looks at my script a bit dubiously:

"I think this should be okay. I will take it with me and prepare a translation."

"Don't lose it, for heaven's sake. That is the only copy I have."

"I won't, don't worry," he says grinning and departs with a roar on his motorbike.

The day's sightseeing now begins. Kaji and I walk to the Durbar (or Palace) Square of Patan where I admire the ancient wood carvings on the temples, and the bronze statues of gods and goddesses. Patan is the second largest town in the valley and is separated from Kathmandu by the Bagmati River. Historically, it is known by its Sanskrit name, *Lalitpur*, or City of Beauty. I am particularly struck by the tall column north of the Harishankar Temple, which is topped by a golden statue of King Yoga Narendra Malla from the 17$^{th}$ Century. He kneels on top of a lotus bud and is protected by the hood of a cobra. On the cobra's head is a bird figure; legend has it that as long as the bird remains there the king may still return to his palace.

From here we taxi to Durbar Square in Kathmandu, where my main interest lies in a visit to the old Royal Palace (now a museum). This is where the city's kings were once crowned and from which they ruled. As such, the square remains the traditional heart of the old town, though the palace was moved north to Narayanhiti over a century ago. We spend some time wandering around the steps of Mayu Deval, a temple to Shiva, and probably the most popular meeting place in the city, where I enjoy watching the fruit and vegetable sellers, the comings and goings of taxis and rickshaws and the souvenir-sellers accosting tourists.

For lunch we go to a rooftop restaurant and take in the views of the square as we enjoy our meal of vegetable chow mein and a glass of Fanta. By now I am understanding Kaji's broken English quite well, and in turn he is growing more confident after a gap of many years.

"Next time you come to Nepal," he says, "I would like to take you to Pokhara and do some trekking around the foothills of the Annapurna Range."

This all sounds very enticing, but I don't anticipate returning to Nepal, let alone trekking in the Himalayas. How wrong can one be!

"It has been going through my mind that I would like to buy Amita a computer," I tell him. "I am sure she needs a computer now for her studies."

"Yes," replies Kaji thoughtfully, "that would be much appreciated. She would certainly find a computer useful."

The next visit is to Budhanilkantha on the northern edge of Kathmandu, to see a famous statue called The Sleeping Vishnu. Our taxi ride takes us out of the city, and at last I get a glimpse of some countryside and can breathe more easily. We travel past fields of golden rice ready for harvesting and watch men and women raking and tying their bundles into sheaves. On arrival, we walk past a few small shops selling postcards and tourist gifts and soon arrive at our destination. But, as a non-Hindu, I am not allowed to enter the sunken pool enclosure where lies the five metre long image of the god. I have to make do with peering over the fence.

Vishnu has many incarnations and in Nepal he often appears as Narayan, the creator of all life. From his navel grew a lotus and from the lotus came Brahma, who in turn created the world. This sculpture was created in the 7th or 8th Century from a single piece of stone somewhere outside the valley, and then laboriously dragged here. Now it lies peacefully on a most unusual bed: the coils of the multi-headed snake Ananta whose eleven hood-heads rise protectively around Narayan's head. It's a curious, if slightly macabre sight.

The return journey is by minibus which is crammed full of people and makes for a colourful if bumpy ride. The 'conductors' on these journeys (for I was to take several more) seem to be boys of about twelve and the drivers not much older. There is a lot of joking and banter on the way. It is the conductor's job to ensure that his minibus is always as full as possible: he whistles, bangs on the door and calls out for more customers at every stage, clutching his wad of rupee notes. He does not seem to worry too much about taking fares, although his passengers offer him a few notes as they alight. The important thing is to maintain a full load, whatever the discomfort inside.

I notice a sign to the British Embassy.

"Kaji," I suddenly call out. "Can we get off here? I would like to visit the Embassy."

I happen to have been given the name of the British Ambassador before I left, and I feel sure he would be delighted to see me. Several security guards later, however, we received the humbling response that the Ambassador's diary is too full for social calls today. I boost myself with the thought that it was his loss, not mine.

The next day is my first teaching day, so I set off on my own in the direction of the Swayambunath Temple. These

walks are my opportunity to watch the local Nepalese at work, well away from the touristy Thamel area, and they provide me with a fascinating collection of photos - the local butcher barbecuing a chicken with a blowtorch; the two elderly gentlemen squatting on the roadside grinding knives; the tailor running up a suit on his Singer sewing machine; the pavement café with its pan of sizzling fat, cooking what look like savoury doughnuts; a small group of men on a bridge sewing mattresses and eiderdowns. All the while, children are walking to school in threes and fours, smartly dressed in their uniforms and carrying satchels – a quaint and rather appealing sight.

Four classes have been assigned to me for lessons at Triple Gem and I feel well prepared. I had been warned in England that I may have to drop any pre-set ideas of how and what to teach, because nothing will be quite as I expect. Brushing these ideas aside, I boldly step into my first class and, after initial introductions, begin talking rather solemnly about nouns and verbs. To my surprise, the children know all their parts of speech perfectly and can even distinguish between verbs of action and verbs of state. Slightly unnerved, I launch into the distinguishing features of common and proper nouns; they have covered that equally well. Hmmm, what now?

I quickly decide to put the ball firmly in their court. Handing out pieces of paper, I ask these children of about nine or ten to write about themselves. What is your name? Where do you live? Tell me about your family. What is your favourite subject? Your favourite food? What sort of person are you? (I am getting into my stride now.) What do you like about Nepal? What do you dislike about Nepal? What is the best thing that ever happened to you? What are your ambitions for the future? And I tell them, "If you don't know what to write, do a little drawing."

I write these questions up on the blackboard – and am met with stunned silence. Perhaps creative writing is not quite in the usual scheme of things at Triple Gem.

"I will give you fifteen minutes to complete one side of writing and then you can read it out to the class."

The brief given to me by Lama Kondan was to be very strict, take no nonsense and give them plenty of English grammar and spellings. But something inside me now reacts against this. I feel they have already been subjected to enough of this approach, so I am trying a different tack.

While they settle down to a period of quiet, concentrated work, I look around. The classroom is small, bare and plain. There are no pictures on the walls, which are whitewashed and somewhat pockmarked. The pupils are sitting at old -fashioned wooden bench desks, the type I remember from my childhood. They even have inkwells! Boys and girls sit separately, perhaps by choice, but all of them are squashed together with little elbow room. They are smartly dressed in a grey uniform with a white shirt but distinguished by a red, green, yellow or blue tie. The teacher has a plain table and chair, and there is a blackboard with chalk. Time ticks by during this period of silent concentration.

When they finish, some of the children choose to read their work out loud while others ask me to do so. Bearing in mind that they are writing in both a language and a script that is non-native, the quality of their English, as well as their handwriting, is astonishingly good. At the end of my first day of teaching, I come home with about a hundred pieces of writing which are among my most treasured possessions from Nepal. Here is one complete piece, typical for its honesty and openness:

*"I am Dristi Shrestha. I am 14. I also have my cute sisters and my lovely mom. She is a nun. And now I am a hostel student. I am interested in visiting new places and meeting new faces. My hobby is to learn new things working with computer and also I play games with my friends. I'm very keen about football. What I like about Nepal is its beauty and love of people. I feel proud by myself that I'm born here and known as Gorkhali. I have nothing that I don't like about my country. But I know there are lots of things wrong about Nepal so I will solve them all when I will be able to stand on my own feet. My biggest wish is to have a good job and look after my family and give them as much happiness as I can, because I know they're in trouble."*

Here are a few sentences taken from other scripts which I found particularly touching. All are copied out verbatim.

*"My name is Karchung Giurung and I am 15 years old. I have no father and mother. When I was 10 years old I was brought by one person who was my father's sister. She kept me in the Orphanage and now I live in Orphanage home...there are 50 children like me...I am an honest girl. The best thing happen to me is that I got a good opportunity to read or go to school and in the future I want to be a nurse and I want to help to the orphanage like me and poor, needy people."*

*"My name is Upesh Dangol. I am a boy of 14 years old... I am jokey, romantic and sensitive person. I don't like the person says that 'they are the one who is best'...I wish to build a hospital in the villages with my money after I read and be business man."*

*"... I am active in all work as well I am well-disciplined. When I was sick and unable to come to school all my friends came to my home to look after me. This was the best thing and best*

*moment of my life. My biggest desires are to become a doctor and go helping poor and needy people."*

*"My name is Bhanu Chandra and I have about 3.5m height. I am 13 years old boy…my favourite things about Nepal are there are different high peaks and green landforms. There are many forests with different kinds of beautiful animals, birds and plants. There are many rivers and lakes. In Nepal there are many historical inscriptions and old monuments which are very beautiful and famous. Therefore I like it. There are different kinds of dislike for me about Nepal as Nepal is not well-developed from other countries. People are starving from hunger and in town area there are different kinds of pollution created by human activities. People are not well jobs so they cannot fulfill their family welly. I am a serious boy…"*

*"My name is Subama Shrestha. I am 12 years old. I live in Swayambhu. My family is small and happy. There are father, mother, brother, sister and me. We never quarrel with each other. We are all very happy with each other. My father is head of my family. My mother is heart of my family. My home is in village. My village very beautiful. I love my village very much. There are more than 4000 people in my village. They are all very polite. They never quarrels with each other. They help each other."*

I find the children here extremely bright and alert – as well as playful and mischievous, given half a chance – so I have to be quick-witted enough to maintain their interest and keep them occupied. In the next day's lessons I use a map of England I have brought with me and point out where I live, where David Beckham plays football (this was 2009, after all) and where the Queen's palaces and castles are. This leads on to a bit of a history lesson, and then we play Riddles which are hugely popular:

*I am round and float in the air*
*But I soon come down to earth.*
*People like me but still they kick me*
*I am hard but can be soft.*
*What am I?*

Football being such a universal topic, this is quickly guessed and they are soon eager to try their hands at writing their own riddles. So pass my first two teaching days and I am adopted warmly into their midst with the rather endearing title "Mr. Neek." At last I am fulfilling my lifelong ambition of teaching abroad, albeit not quite as I imagined.

# Chapter Three

It is Saturday, the day when the Nepalese Christians traditionally go to church. Shyam Nepali arrives on his motorbike at 8.00 am and I hop on the back for my lift to Patan. The streets of Kathmandu are crowded and noisy, and I find it a somewhat hair-raising experience, so I clutch the pastor firmly round his waist. But I am in safe hands and after a twenty minute ride we turn off down a sandy, unmade side track, strewn with heaps of litter and black refuse bags, and arrive at Golgotha Church. It is a building which, from the outside, resembles a long double garage. I am taken inside and invited to sit on the only chair at the back of the building. Gradually a few people drift in and take their places on the floor - women on the left, men on the right. Amit soon arrives on his motorbike. He is wearing a smart suit and carries a guitar case.

"Hi, Nick. Here is your writing. When my father calls you to the front, I will stand beside you and translate. It will be fine," he says encouragingly.

The church is totally plain apart from an altar with a simple cross and a lectern where the pastor will stand. There is a gentle humming from two ceiling fans, and three metal-framed windows are open to help keep the room cool. The walls are a pale yellow colour, peeling in places, but there is no adornment of any sort. Outside, leaves from the trees in the garden brush against the window panes and sunshine streams through, casting long shadows across the floor. Amit now starts to play and sing songs, continuing for a whole hour, his eyes closed in devotion. These, I learn later, are Christian devotional songs which he and his friend Radheshyam have composed. Meanwhile, the church gradually fills up with both young and old, and there are several babes in arms. Other musicians arrive and join Amit on the platform: a keyboard player, a percussionist and another young man as vocalist. I take a picture of this group, which will have an interesting significance for me later on.

The service starts at about 10 o'clock and for the next two hours I really do not understand anything, but am happy to be sitting there listening and observing, joining in the songs as best I can. At one point, Shyam's assistant pastor whips up the congregation into a frenzy of chanting, waving and clapping. I watch from my seat at the back, where I am a little embarrassed to find myself sitting on the female side, but nobody seems to mind. There now begins a period which I can only describe as 'speaking in tongues'. Everyone seems to be saying their own prayers out loud, praying for themselves and others. This period, I later gather, is called "Worship".

Soon there is an announcement and I assume that I am being introduced because people turn to look in my direction; my turn has come. I have brought along the well-known story of *Two Frogs in Trouble* and a frog glove puppet. I had planned to use this at Triple Gem School but like some other of my ideas I quickly dropped it as being too juvenile and inappropriate. However, all congregations like a story with a strong moral, so this is what I relate to these good people, with Amit's help:

"One day, Little Frog said to Big Frog, 'Come on, let's go and have some fun in the farmyard.' Down at the farm it is milking time for the cows and with all their hopping, the two frogs end up jumping into a big pail of milk. What a disaster! They swim and swim, but no one hears their cries for help. 'I can't keep this up,' says Big Frog, but Little Frog calls back, 'Keep going, keep going! Never give up hope!' But Big Frog cannot keep up his paddling and eventually disappears beneath the surface. Little Frog, however, thinks to himself, 'To give up is to die, so I will keep on swimming.' Two more hours pass and the tiny legs of the determined frog are almost paralysed with exhaustion. But when he recalls the fate of Big Frog he thinks, 'I will keep on paddling till I die – if death is to come – but I will not cease trying. *While there is life, there's hope.*' So he keeps swimming and swimming, chopping the milk into little waves. After a while he begins to realise that the milk is becoming thicker. First, it turns to cream and soon he notices that he is resting on a small lump of butter that he has churned by his incessant paddling. And so the successful frog is able to jump out of the milk pail to freedom. 'Never give up hope,' Little Frog calls out to the cows as he goes hopping off on his way."

The story seems to have kept the congregation spellbound, and I suspect they have never been told a story quite like this in church before. There have been some ripples of laughter, so I think it has gone down quite well. I walk back to my seat, smiling at the bemused stares of the children as I step through them, and I can now relax for the rest of the service.

Towards midday, a healing service takes place which consists of the two ministers laying their hands on particular members of the congregation, while everyone else holds their arms up in the air and gives vent to loud chanting. It is altogether a noisier and more energetic affair than I am used to in England. One elderly lady, who is badly disabled and uses a zimmer frame, seems to benefit from this service, so words of praise and general thanksgiving are offered up. This brings the service to a conclusion. Later I have a chat with the assistant pastor who describes the particular style of worship here as 'evangelical charismatic' – a pretty fair description.

After some friendly greetings from the congregation, I am given a ride back to Amit's home and 'recording studio', where I have been invited to have lunch with his family. The home is, I am told, typical accommodation for a young couple. It is on the edge of some allotments and consists of a downstairs room with a kitchen area and a small upstairs room with a bed, a few chairs and some recording equipment. It is all very dark and pokey, and the two floors are connected by a ladder. This is where Amit and Radheshyam write and record their evangelical songs – and from what I can understand, this is their daily occupation. I am given a copy of their latest CD and invited to make a contribution to the cost of the next one. I gather that there is an ongoing commitment to record all the hymns in their church hymnbook and distribute them to the Christian communities in the outlying rural areas which fall under the Golgotha parish. (One day I will be visiting these communities, but I don't know this just yet.)

I am offered tea and biscuits and, in a while, a meal of rice, *dal* (lentil stew), raw white carrot, spinach and a very strong pickle. The men sit upstairs, served by Amit's wife, but she has her meal with a couple of other ladies downstairs. Even though conversation is slightly limited by the language barrier, the atmosphere is friendly and calm. They seem pleased to have me as their guest, and I suspect it is the first time they have welcomed an English person into their home.

Towards the end of the meal Shyam turns to me and asks,

"So I will come and pick you up again next Saturday at the same time, and perhaps you will speak to us again?"

"Of course. Thank you for asking me," I say without a moment's hesitation. This will give me a week to think of something else to say.

After lunch I descend the ladder into the kitchen, thank the ladies for the meal and accept a ride to Durbar Square, from where I will easily be able to walk back to my hotel. I set off by myself with a pleasant sense of feeling at home here. Having been invited to share a meal with a Nepalese family, I feel accepted into the community, especially since I have been invited to see them again next week. To feel at home in Kathmandu… how strange is that? Maybe I *will* be returning here one day.

With my *Lonely Planet* guide book, I take a self-guided walk and enjoy some sightseeing before returning to my hotel. Reflecting on the morning's service, I suspect there is considerable poverty in the congregation and yet a strong sense of devotion as well. I had not really expected to encounter a Nepalese Christian community in Kathmandu, but this church is clearly thriving – with a good spread of ages as well.

The next day I meet up with Kaji who takes me on a bus ride to Nagarkot (about twenty-five miles east of Kathmandu) where we hope to see the sunrise over the Himalayas. The first part of the journey is an hour-long dusty ride to Bhaktapur where we have lunch at a small café. The meal of fried vegetable noodles is served in a palm-leaf bowl, and it is as cheap as it is delicious. As we take a leisurely walk around the town, I see local women sieving rice to extract dust. There are several old wooden houses built in the traditional Nepalese style to admire. It is a relief to be away from the noise and pollution of Kathmandu. Bhaktapur is the third major town of the valley – and really warrants a more extended visit – but our bus to Nagarkot is waiting, so we climb aboard. Before too long it is full, but this does not deter eight young men from climbing onto the roof. I understand they still have to pay the full fare.

The two-hour drive is a steady ascent of twisting roads through terraced rice fields, harvested by ladies in colourful saris. Sometimes the countryside takes on an almost alpine look, as we pass through pine forests and lush green vegetation. When we arrive in Nagarkot, I begin to wonder where we are going to stay, but I need not have worried because a man approaches us down a dusty track.

"Are you looking for rooms?" he asks.

Yes, we say, so he leads us back up the track to an attractive hillside hotel called Hotel Snowman, with wonderful views over the surrounding countryside and eastwards towards the Himalayas.

"How much?" I ask him. (I am a little concerned about my cash, because I am having problems taking money out of ATM machines.)

"800 rupees per room, with breakfast," he replies.

That's about £6.50 each, which is amazingly cheap and I can just afford it, so I count out the money and we are settled for our night away. After showering, we meet up on the hotel terrace for an evening drink. Now Kaji begins to tell me a little more about the circumstances which led to the end of his career as a guide.

"I was leading a party of trekkers on a trip I had done several times before. But all of a sudden one day, when we had reached the highest part of the trek, I was overcome with feelings of nervousness and tension. I had never felt like this before. Then I blacked out. A rescue helicopter had to be called to airlift me to hospital where I was told I had pneumonia. My employers laid me off with no compensation or sick pay. This was two years ago. I have been offered the post of assistant guide, but after being leader this did not interest me – so now I have no proper paid work."

Kaji is thinking of setting up a small business at home, perhaps selling vegetables. He shows me on a map where his family lives. It's an area called Dhapasi just north of the city and about forty-five minutes' walk from Thamel.

"We moved there from the city ten years ago, when it was mostly farm land, but now it is very built up. Come and visit us, and I will show you around."

As the afternoon light is beginning to fade, we walk up to a high vantage point in the village, from which we can see the snowy peaks of the Himalayas glowing pink in the setting sun. It's a wonderful sight, but he tells me it will be even more spectacular at sunrise the following morning. Returning to the hotel we have a cup of tea together, but

I don't feel like supper, so I leave him to enjoy a meal and watch TV in the hotel restaurant while I go to bed for an early night.

I am interested in the fact that I have brought a heavy rucksack for my overnight stay, while Kaji appears to have brought nothing at all. And apparently all showers are taken cold in Nepal – so in winter they limit themselves to one a week.

I am up at 4:00 am in the hope of seeing a spectacular sunrise, but I am likely to be disappointed as there appears to be thick heavy cloud in every direction. I quickly throw on some clothes and, leaving our hotel, walk up the road to where the more expensive hotels are situated in the hope of finding a better vantage point. I spot one smart establishment which has a high viewing platform, so I surreptitiously make my way there and join some Japanese visitors who have lined up their expensive cameras on tripods. It is obvious we are out of luck, but they take pictures anyway of the dull grey sky. I walk back to Hotel Snowman from where Kaji is just emerging. We smile a bit ruefully to each other, pointing to the clouds overhead.

"Oh well," he says, "you will just have to come again another year."

Once the disappointment of the sunrise is behind me, I am able to enjoy whatever else the morning brings. Our bus isn't due for another two hours, so I go for a short walk through the village after breakfast. At a local school, children are gathering in the playground to play football. I notice the sign: "Rotary School – dedicated to the education of young minds". Further down the tarmac road about thirty soldiers in full uniform are running uphill in close formation towards me, some carrying rifles. As they approach and run past me, I reflect on the close links between the British

Army and the Army of Gorkha, as it was originally called and from which the name 'Gurkha' was derived. To serve in the army is a highly-prized honour in Nepal, so these soldiers will command considerable respect in their society. I watch them disappear into the distance, then amble down the hill to the bus stop area. The clouds are clearing now to reveal a bright blue sky. There is a magical stillness in the air, broken only by the sound of people emerging from their homes to wash their face, brush their teeth or collect milk from large roadside urns. There is a plateau of level ground a little further off where I see young boys in judo outfits having some early morning instruction. The Kathmandu Valley appears lush and fertile in the growing light, and I notice the traditional terracing of nearby farmland abundant with potato crops. Soon the silence of the early morning is broken by the noisy arrival of our bus, and we climb aboard for the homeward journey. We are back in Kathmandu by midday and I thank Kaji for the trip.

"Come to our house next Sunday for lunch," he says. "I would like you to meet the rest of my family."

I thank him, we shake hands and part company. Now my main concern is to find an ATM machine where I can take out some cash. I try one after another but without success. So I send an email to my partner, Adrienne, at home in England and ask her to get in touch with my bank. I suspect they have put a block on my credit and debit cards The next thing I have to do is pay my hotel bill (where they only take cash), and find a different hotel where I can use my MasterCard.

All this takes until late afternoon, by which time I have moved into the Hotel Tradition in the Thamel area, where the room costs only 960 rupees a night (which is less than £10). I go to my local internet café and Adrienne has already replied to my email. I must telephone my bank (which has indeed put a block on my bank cards suspecting that they have fallen into someone else's hands). Fortunately the internet café has a telephone facility, and within a few minutes I have got through. I explain that I am in Kathmandu and need to take out some cash. Why I didn't inform them of my trip before I left I don't know – it is hardly surprising

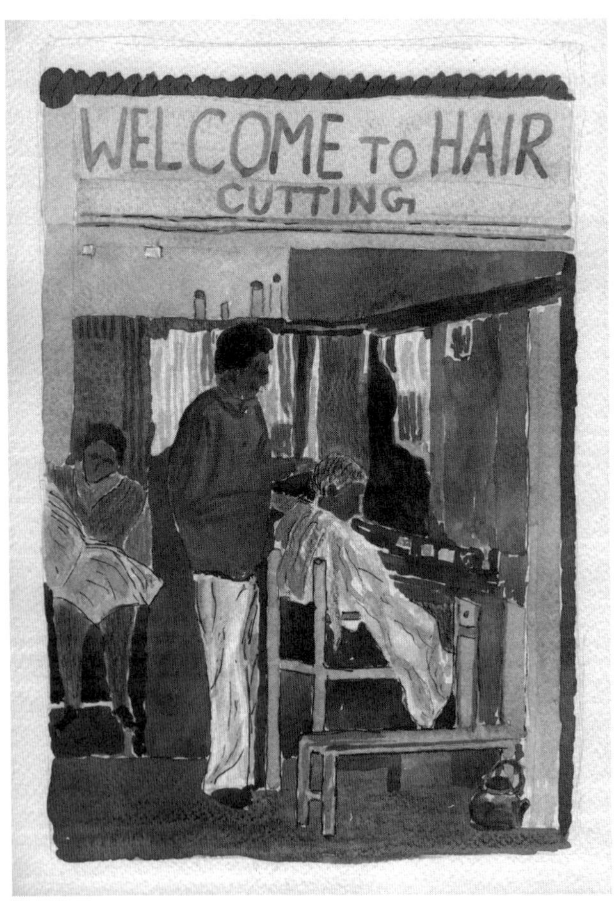

that they were suspicious. Anyway, all is well. They accept my story and will unblock the cards. With a huge sigh of relief, I go and rest in my new hotel room. Global communication is a wonderful thing!

The next morning I set off in buoyant mood for my final day of teaching. School does not start till 10 o'clock, so I am able to enjoy a leisurely walk. My attention is drawn to a large sign straddling a somewhat rickety shop: 'Wel-Come To Haircutting' so I decide to call in. It is a tiny establishment, open to the street, with a couple of plain wooden chairs for customers and a floor covered in as much hair as dust from the outside world. I am warmly welcomed in by the barber who shows me to a chair. Now at home I am accustomed to a haircut lasting ten minutes, but this one lasts the best part of an hour. After my hair has been cut with great care and precision, my scalp is given a very thorough massage. The barber's hands now start kneading my face and eyelids, firmly but gently. Next he indicates that I must lower my head to the table so that he can work on my neck and shoulders. Finally my arms are massaged and my fingers stretched and pulled in a most delightful way. I sit up slightly bemused and see his face glowing with pride, knowing how much he has pleased his Western visitor. I ask him, "How much?" and he shakes his head from side to side with a great grin, suggesting, "Whatever you like!" I think to myself, 'Well, I pay £10 for a haircut at home – but this has been something else altogether.' So I give him the equivalent of £10 which is probably the price of about ten haircuts. But I was delighted – and so, of course, was he. We smile, bow, shake hands and I go on my way.

I have a double lesson with the top class at Triple Gem today and realise once again how bright these young students are. I want to do something which I feel will be beyond their own teacher's range of knowledge and quite outside their curriculum. In the computer room at my hotel I have been lucky enough to find a dusty old copy of *As You Like It* – I cannot imagine what it's doing there, but it has given me the idea to tell them a little about Shakespeare and give a brief account of the play. I will then write on the board Jacques' famous speech, "All the world's a stage..."

This proves to be surprisingly successful. I am amazed at the ability of 14 and 15 year old Nepalese pupils to grasp not only the difficulties of the language, but also the concept of blank verse and iambic pentameter. Again, I am struck by how basic their classroom is: twenty teenage students crammed into a tiny room, sitting at old-fashioned desks on a bare concrete floor, walls painted roughly in yellow and blue, a thin tatty curtain, one pane of glass missing and the only wall decoration a school calendar. But the pupils are smartly dressed, bright and alert, and will laugh uproariously at the slightest provocation. (I have to explain why I am a bit late, and they find the story of my haircut utterly hilarious.) These are delightful kids and seem genuinely happy.

Life in Nepal, I am beginning to realise, is very much shoulder to shoulder, whether on the roads or in buses or in the classroom. It is altogether a shared experience and sometimes shared as much with hens, geese and dogs as with other people. Children squashed together at their desks may seem uncomfortable to me, but maybe it has a bonding effect. No surprise then that the native greeting to each other in the street is 'dai' or 'didi' ('brother' or 'sister') suggesting 'we are all one as brothers or sisters, so let's recognise it and enjoy our togetherness'.

My final teaching day comes to an end and some of the girls have written me little farewell poems. Not all my classes have been an unqualified success, and one in particular descended into outright chaos. But it has been the experience I wanted – just a shame I did not see the boy I am sponsoring. However, I will continue to be sent reports on his progress. On my last day, I bid farewell to Lama Kondan and thank him for the opportunity to teach at his school. He gives me a big hug in return.

"Come again!" he says with a beaming smile. "You can see how much the children love you."

I decide to walk back across some rice fields, a route I have taken once or twice before. I feel particularly light-hearted now that I have finished teaching. It has been surprisingly challenging, and I would not want to be going back for more. So I walk in a very leisurely way this afternoon, enjoying my amble through the fields, watching men and women at work on their allotments, harvesting rice and attending to the irrigation channels where cabbages are growing. I fall into step and then conversation with a middle-aged man called Dr Laxman Shakya. He asks me who I am and what I am doing in Kathmandu. When I mention Triple Gem School, he becomes quite excited.

"Ah, Lama Kondan. A very dear friend of mine."

He explains that he was a Buddhist monk himself many years ago. Now he works for an organisation called Endure, which is the Nepalese branch of World Without Anger (WWA). It originated in the USA and runs programmes for schools on its teaching.

"Have you been to Triple Gem School yet?" I ask him.

"Not yet, but now you have sown a seed in my mind!" he replies with a twinkle in his eye.

# Chapter Four

The next few days pass interestingly and enjoyably but now, with my teaching behind me, I am beginning to look forward to returning home early next week. Several people have been asking me when I will come back to Nepal, that I must stay longer next time and visit such and such a place. But I really have no intention of coming back. Apart from the disappointment of not seeing Manish, it is a mission accomplished as far as I am concerned. I believe that my short adventure is nearing its end - how wrong I am!

I wake up on Saturday 31st October with the wish that I had brought some foot cream with me. After walking many miles through the dusty streets of Kathmandu in a pair of sandals, my heels are now beginning to feel like coarse old sandpaper. However, I have to put that thought aside because Pastor Shyam Nepali will soon be here to take me for my second visit to Golgotha Church. He arrives, helmeted, on his motorbike promptly at 8:30 am.

"My wife Roopa would like you to have these. She made them herself." He hands me two little orange crosses made out of tiny beads. I say I will look forward to thanking her later.

The morning follows the same pattern as the previous Saturday: Amit singing to his guitar, like a call to worship;

people drifting in, and this time giving me a warm greeting; the service during which I am again called upon to speak (my diary entry records that my story was called *The Mustard Seed and the Caterpillar* – which sounds intriguing but I cannot for the life of me remember what it was about); the same young men playing or singing on the platform at the front – and of course everything in Nepalese, so I don't understand a single word.

After the service, an elegant smiling lady wearing a sari and headscarf asks to be introduced to me.

"This is Indra Maya Shahi," says the Pastor.

We bow to each other, and I think I remember her from the congregation last Saturday. She says a few more words to him.

"She would like you to visit her orphanage."

'Oh dear, this spells trouble,' was my instant thought, but I quickly dismiss it and say, "Thank you very much. I would love to come."

Shyam translates, and the lady bows, smiling sweetly to me. We arrange that Shyam will pick me up from my hotel the following morning.

After the service I am again given lunch at Amit's house. This time I find myself talking to his younger brother, Kishan, who is a science teacher. He tells me how ten years ago there was a Maoist uprising in Nepal, and the rebellion had the effect of terrifying the villagers who fled into Kathmandu in their panic. The result was the overpopulation of the city, leading to the congestion and pollution which I have been witnessing. Poor urban planning has only made matters worse, and there is a growing unemployment problem with far too many people chasing too few jobs.

I also talk to Chittra, Shyam's assistant pastor. Was he born into a Christian family, I ask him, or has he converted from Hinduism, or Buddhism?

"No," he tells me, "I am the son of a witch doctor. I learnt about Christianity when I was a teenager, and very soon after that I converted." Last year he visited three of the churches in his parish to perform baptisms. These were all within a 300 mile radius, so it took him two weeks to cover the ground – which he did mostly on foot. The baptisms involve total immersion in mountain streams.

"I would like you to come with me when you visit next time. My parishioners will give you a very warm welcome."

I thank him, and say I would enjoy that very much. After lunch I am again given a lift on Pastor Shyam's motorbike, and he drops me off at Durbar Square as before.

I spend the afternoon by myself, and the time passes uneventfully. I have been reading Bill Bryson's *Short History of Nearly Everything* which has been utterly absorbing, so I continue with that for a while, sitting in a shady spot in Durbar Square. A young Australian couple I had met the previous day come and sit with me. They have just cycled to Nagarkot and back all in one day – staggering!

Later I go to my favourite café in Thamel where I have fried *momos* (a popular dish in Nepal but of Tibetan origin, consisting of meat or vegetables wrapped in a dough, then steamed or fried) followed by yoghurt cake and *lassi*, a yoghurt drink. I spend a little time wandering around the shops, buying small gifts to take back home and taking a few more pictures. There are two events arranged for the following day – the trip to the orphanage and then the visit to Kaji's family – so I am glad my last day is planned out. After that, the prospect of returning home will be very welcome.

So it is on Sunday 1st November, a day of unforgettable significance, that Pastor Shyam arrives on his motorbike to take me to Indra Maya Shahi's orphanage.

After about half an hour of driving on busy thoroughfares, we suddenly shoot off to the right down one dusty side street, then another, equally rocky and dusty, before ending up in a fairly desolate area where stands a two-storey house (the ground floor, I am told, comprises the orphanage). We dismount here and Shyam leads me through an iron gate into a tiny courtyard where onions and garlic are growing by the steps leading up to the front door. Indra Maya must have heard us coming, because she comes out straightaway and with a big smile beckons us in. Two young men are in the house - one seems to limp quite badly and gives me the impression of being inarticulate, while the other seems slightly familiar to me. Didn't I see him on the platform at the church? He has black swept-back hair, dark oval eyes, is slightly built and wears a serious expression. He is introduced to me as Ramesh.

We sit down together in the living room. Indra Maya cannot speak any English, so she indicates to Ramesh to make some conversation.

"We are very pleased to have you here," he says with great seriousness, "and Mama would like to offer you a cup of tea." He speaks good English and is easy to understand.

"I would like that very much," I reply.

There is a long pause. The other boy, whose name I gather is Pancha, leaves the room with Indra Maya. Shyam, Ramesh and I sit silently together. I feel I ought to say something.

"Is this an orphanage?" I ask, rather stupidly.

"Yes," says Ramesh. "It is called Wasta Care Centre, and there are five orphans living here, and I am the second oldest. The other four are away today."

"So who is Pancha?" I asked.

"That is Mama's own son. She has four other sons and a daughter. When they grew up, she left the house to them and opened this orphanage. But she brought Pancha with her because he needs special care."

I ask if he would write down the names of the five orphans, and this is what he enters with great care into my diary: Niran (20), Ramesh (19), Anish (17), Sunil (17), Arun (13). He hands me back my diary and there is more silence while we wait for the tea.

"Are you a student?" I ask.

"No, I graduated from High School two years ago. Niran and I both went to college to complete Grade 11 and 12. I graduated in science and he graduated in commerce."

"What have you been doing since then?"

"It has been very difficult. We have done some Sunday School teaching, but really we have not been able to get proper jobs."

Indra Maya comes in with the tea, and says something to Shyam, who then turns to me.

"Indra Maya has been very worried about Ramesh and does not know what to do."

I turn to the young man and ask him how he spends his time.

"Well, really I am wasting my time here. I try to study – look, do you see?" and he shows me some maths and physics text books, "but I cannot do anything worthwhile. I really want to be a doctor, but no college will have me because I don't have any money – so I am staying here, not really doing anything."

He looks so downcast and crestfallen that I hardly know what to say next. I begin to realise why I have been asked to come here: perhaps I can come to the rescue. I hesitate in the long silence that follows, knowing that an offer of help from me could give him some hope. But what would I be getting myself into? I wait, take another sip of tea, then say, "How long have you been living here? Tell me a bit about yourself."

"I was born in a poor family in Sindhupalchok, Langhkhola, a rural area in the northeast. I do not remember the face of my father, who died when I was three or four years old. After the death of my father, my mother got married with another man from the same village and we went to the Terai area in south Nepal. I have in total three brothers and sister. One brother is older than me and he has gone abroad and we have not heard about him since then. He went abroad about four years ago. He wanted to go to Malaysia but since then I have heard he is in Singapore working as a security guard. But I have not heard anything from him. My other two brothers are younger than me. One is impaired with a hearing problem and the other brother is staying with one uncle and having problems about studying. He wants to study but he does not have the support he needs.

"My sister is the youngest and she is with one relative. She stays there and works. In the daytime she goes to school. She works and studies there. And she is really good at study. She was the first in every class. But nowadays I have not heard from her. So I do not know how she is or what she is doing. You know, when I remember each of my family members, I feel like crying because they are still having problems in their lives and I am enjoying a good life in this orphanage. I pray for them and one day they will be pulled up, I know."

He pauses to drink some tea. I ask him if he would tell me what he remembers of his childhood.

"When I was so young, I had to work in the brick factory with my step-father. Most of the time I used to work in the factory and I did not know about school. I used to wake with him and go early in the morning. You know, my step-father was a strict man and he would beat me for simple mistakes as well. And I have a guess that my current ear problem may be due to that beating. He used to beat me wherever he finds me. I was the oldest of all, as my eldest brother was with my aunt in Kathmandu. So my step-father would beat me even for mistakes of my younger brothers. And he used to drink and would fight with my mother as well. I used to cry and just hold mom to avoid them fighting. But they never stopped fighting. And sometimes, he used to beat me for supporting my mom. And I was living with fear every time. You know, I used to be scared sometimes even in my sleep. One day, he beat me and nearly my urinary bladder was injured, and I had blood in my urine. So, I had a really tough time with him as well. Later when I was around nine years, he died of tuberculosis. I do not know why he never showed me love but he loved my own younger brother more because he was weak and unhealthy from his childhood.

That may be a reason. Also I think he was not so well with his brain as he was suffering from different diseases in his childhood. I used to work in the morning with him and when I was back from school, I used to carry bricks on my head. I used to carry them with my mother and that might be a reason for my back pain which led to a disc prolapse. Mom used to cry for me and she wanted me to study but step-dad wants me to work first and if time was left go to school. You know, I had to support my family because he was neither a healthy nor an understanding step-father. When he died, then my schooldays were totally ruined and I had to be with my mom all the time. We had loans to pay which were taken for the treatment of step-father. He was sick nearly for one year. He cried before he died. I was so scared with him, and I could say nothing. Whatever I needed, I always asked with my mom. Then, after his death, mom also started to drink for avoiding tension. That was the most hard time of my life. I was only nine years old and whole family was in grief. I did not know how it feels to lose a husband, but I could see it in the eyes of my mom, the tears and sorrow. I crushed myself that time for being born in that poor and homeless family."

I am very moved listening to his story and amazed at his fluency and the extent of his vocabulary. Where did he learn the phrase 'disc prolapse' from? But all the time I have been listening, I have also been giving his situation some thought. I am conscious of the fact that my parents paid for my private education and that of my two brothers but this was a responsibility which never came my way, as I did not have children of my own. Should I perhaps offer to pay for the further education of this young man? One part of me wants to have nothing to do with it – especially on my last day in Nepal – but the other part knows that if I don't help, who will?

"Is there anything I can do to help?"

Ramesh lifts his head, looks straight into my eyes and says, "You can pray for me."

This isn't quite the answer I expect, and rather falteringly I say, "Yes, of course I will pray for you."

There is something about this boy that is speaking out to me, for despite his downcast air and sad expression I feel there is a strong character and real sincerity here. So I hesitantly press on:

"Is it possible to do medical training in Nepal? And is it very expensive?"

"Yes. There is a college in Kathmandu, where you can train to be Bachelor in Medicine and Surgery. But it is very expensive."

"How much does it cost?"

He writes down again in my diary a huge sum of Nepalese rupees, which I gather is the equivalent of about £40,000. I gulp slightly at this but go on, wishing I suppose to keep the boy's hopes alive, even if only while I am in his company. I take a deep breath and say:

"If you manage to obtain a place, there is a possibility that I might be able to support you."

He does not show much reaction to my comment but maintains his composure. Shyam talks to Ramesh and Indra Maya at this point, and I am left out of the conversation, which goes on for a few minutes.

Eventually Shyam says, "They would like to take you to see Kathmandu Medical College. Two of Indra Maya's sons will come to your hotel with Ramesh tomorrow and take you there."

"All right," I reply, and tell them I am staying at the Hotel Tradition in Thamel, and will be standing outside the front door at 10:00 am. Once this arrangement is made, there is not much else to say. Indra Maya smiles and bows to me, but Ramesh continues to look solemn and downcast; perhaps his life as an orphan has left him so embittered and disappointed that he cannot believe that any good fortune might come his way. But really I have no way of knowing. We part company, and Shyam drives me back to Thamel. I hope he feels pleased with the way the meeting in the orphanage has turned out; but for my part I cannot manage to still the seeds of doubt that are germinating in my mind.

I try not to think about it too much and am determined to enjoy the rest of the day by myself. Back in Thamel, I buy Adrienne a Tiger Eye necklace from a street-seller, and in an art shop bargain for four small watercolour paintings, which I eventually buy. I go to my familiar restaurant for some supper before returning to the hotel to read Bill Bryson. I am tired by about 10:30 and decide to turn in.

# Chapter Five

I now realise why my hotel room is so cheap. It directly overlooks the noisy streets of Thamel – and on this particular evening, when I have a lot on my mind, I find for the first time that I cannot get to sleep. The shopkeepers keep lifting and lowering their doorway shutters, making them sound like the pounding of metallic waves on the seashore; motorbikes constantly drive past, beeping incessantly; from somewhere across the rooftops comes the endless blast of piped music; dogs howl in pain and distress… and above all I have a splitting headache and a sore throat. It is a very warm night but I have to switch off the air-conditioning because it is so noisy. This means keeping the windows open and putting up with all the outside noise.

There is no possibility of sleep, and as the hours tick by, my seeds of doubt are growing into a mountain of worry. Suppose the whole of today's meeting has been one big set-up? Who are these two sons of Indra Maya who are going to drive me off to some unknown destination (unknown at least to me)? I am due to fly home early on Tuesday morning but now, at the eleventh hour, something is going to detain me. In fact, I am possibly going to be kidnapped!

As absurd as all this now seems, during that particular night I am totally possessed by this train of thought. Sleep is out of the question. The noise from outside and the turmoil in my mind show no signs of abating, so I must take action. I get up and start packing. I have already paid my hotel bill, so I decide to leave Kathmandu a day early. I will take a taxi to the airport and catch an earlier flight home. Once packed, I write two notes: one to Ramesh and one to Kaji (who has invited me to visit his family the coming afternoon). I apologise for not seeing them as arranged today but I am ill (partly true) and have to fly home. I leave Ramesh my email address, telling him to send any information about his course to me at home and not to give up hope.

I take my luggage down to the lobby and wake up the night porters. I ask them to order me a taxi and leave the notes for Kaji and Ramesh at the hotel desk. The taxi duly arrives – and by 4:30 am I am at the airport. It is closed, and the armed guard will let no one in until it officially opens at 5:30. So, here I find myself standing forlornly outside Kathmandu Airport, hoping to board the 9:00 am flight to Doha, for which I do not even have a ticket. My ticket is for the flight at 9 o'clock the following morning!

In the cool light of dawn, the absurdity of the situation I have landed myself in starts to become apparent. There is practically no chance I will be able to switch flights at the last minute, and I cannot possibly wait at the airport all day for tomorrow's flight. So I will just have to find another hotel for one night. I cannot bring myself to return to the horrors of Thamel, and the only hotel I know of in a quiet area is Hotel Greenwich Village in Patan. So as soon as a taxi appears I will ask to be taken there.

I wait and wait – and now of course I begin to realise that I was never going to be kidnapped, and the whole thing has been a ghastly mistake. And then I think how disappointed Ramesh and Kaji will be in a few hours' time to read my rather pathetic notes and think that I have done a bunk. How feeble it seems now to contemplate leaving a day early with my trip incomplete.

A taxi draws up, letting out a passenger – I hesitate and let it go. Maybe I can just catch this morning's flight to Doha. Soon two more taxis draw up, but still I don't move. Several passengers climb out, gather up their luggage and walk towards the terminal building. The taxi drivers stand there, chatting and smoking. In the dawning light, I emerge from the shadows with my luggage.

"Can you take me to Hotel Greenwich Village, please?" I ask one of them.

They look at each other, then one of them asks me something but I cannot make it out. I say simply, "My flight has been cancelled. I need to go to a hotel."

He loads my bags and we set off. With luck, I can just about retrieve the situation. But there is one further setback because, when we arrive, the hotel turns out to be full. The receptionist suggests I try the Summit Hotel, so I duly lug my suitcase there; it is not too long a walk, fortunately. The Summit (which looked even more expensive than the Greenwich) can offer me a room, but it will cost me about £100. By this stage I have no option and take it without a murmur. At least I will have a good night's sleep, and a taxi will come at 6:30 the following morning to take me to the airport.

I hand over my MasterCard and the room is booked. So simple! I am shown to my room and can hardly believe that such luxury exists in Kathmandu: a lovely big bed with bath, complimentary bowl of fruit, quiet location and clean air. How fortunate I am that I can simply walk into this hotel and afford to pay for such luxury! I wonder what Pastor Shyam Nepal and his congregation would think of my night's exploits.

But I have no time to linger, because I must now set off on the long walk back to Hotel Tradition. By now I am curiously excited about the prospects of the coming day – and despite feeling very groggy through lack of sleep, I enjoy my early morning walk through the quiet streets of the city. Perhaps the prospect of my final night at the Summit Hotel is buoying me up.

I arrive just over an hour later. Thamel is beginning to wake up and the hotel night porters have left. I ask the receptionist for my room key (the room is still mine until tomorrow morning) and for the two letters which I had left here earlier. She hands me all these as though nothing has happened and I go upstairs to my now empty bedroom. I no longer feel the least bit tired, but am instead ravenously hungry. I go up to the Terrace Restaurant where I am served a delicious breakfast of mango juice, scrambled eggs, tomatoes, hash browns, tea and toast. The chilliness of my early morning walk has by now turned into a warm, bright morning. As I stand by the terrace balustrade looking out over the sunlit city, I reflect again on my good fortune.

I would have had to change hotels anyway because another night here would have been unendurable. I suppose I needn't have departed quite so hurriedly, but then I was in a considerable state of panic. Yes, it is going to cost me a lot of money, but for the peace of mind and a good night's rest (and a warm bath), I feel it is worth the cost. So I no longer feel too mortified by the turn of events; in fact, considerably relieved and calm. I am not after all going to be kidnapped – what tricks the hours of night play on our vulnerable minds! – and above all I am glad at the prospect of spending a few hours with Kaji's family this afternoon.

I pay for my breakfast, go down to reception where I hand in my key for the final time and stand outside the

hotel entrance to await the arrival of Ramesh and his two brothers. To be honest, I am fairly indifferent about whether they turn up or not. Half of me still wants to honour my offer of help to the young man, while the other half wants to have nothing further to do with him. I stand in the sunshine watching the shopkeepers push up the noisy metal door-shutters which had so disturbed me last night. A pair of elderly men approach with some curious equipment which intrigues me. They settle down on the dusty street and start to ply their trade – knife-sharpening. A passing shoeshine boy waylays me but I am not interested.

"Later perhaps," I say.

"Thank you, Papa," and off he goes.

It is by now ten past ten, and I decide to wait until half-past. That will at least allow thirty minutes delay for our rendezvous. I am already beginning to compose the note (the second note) which I am going to leave at reception. It will go something like this:

"Dear Ramesh, I waited outside the hotel for you and your brothers to turn up, but by half-past ten you still hadn't arrived so I thought perhaps you had changed your mind – or something had occurred to delay you – but I had no way of getting in touch. Here is my email address so please write; I am sure I will be able to help you. Best wishes, Nick Morrice."

I have to admit that, after a good half-hour's wait, I am relieved to be able to leave this note at reception and walk away into obscurity. Looking back, I could easily have waited in the lobby for the rest of the morning for him to arrive, knowing perfectly well what the traffic delays can be like in Kathmandu, and I had nothing else planned. But I am sorry to say I just wanted to walk away from a potentially difficult, complicated and expensive situation so this is what I did. But at least he will have my email address if he does arrive.

I drift into Thamel's rabbit warren of streets and soon catch up with the young man who waylaid me earlier.

"Shoeshine now, Papa?"

"Well, possibly. How much is it?"

"75 rupees each shoe, Papa."

"Too much," I say, and walk on.

But he runs after me.

"100 rupees for both shoes, Papa."

"OK." I will enjoy having my dusty shoes polished for my one night in the Summit Hotel. I settle down into his portable chair which he has set up directly in front of a small clothes shop in a relatively quiet, traffic-free part of Thamel. While he kneels at my feet and starts brushing I enjoy watching the passers-by and soon attract the attention of a young couple from America. They tell me about their plans to go trekking in Pokhara. After they have left, I ask the shoe-shine boy his name, and he tells me it is Om. I tell him mine but he doesn't seem too impressed and continues to call me Papa. It turns out that he is a student of Buddhist Thanka painting.

"I will take you to my school, Papa."

"Oh, really?"

"Yes, Papa. It is only a short distance from here. You will enjoy it very much."

"Fine, but I must be back here in an hour."

"Don't you worry, Papa, I will look after you."

What am I letting myself in for this time? Om has made an excellent job of my shoes, and he is by now so excited that he is soon on his feet, gathers up his bits and pieces and is skipping ahead through the crowds, beckoning me on. I hurry as best as I can, just keeping him in sight, but the streets are thronged with people and Om is quicker and more nimble than I am. Thamel is such a tortuous maze of twists and turns that I have to keep careful note of the route as we scurry along so I will be able to find my way back. Fortunately there are plenty of signs in English to help me out.

We eventually arrive at Om's School of Buddhist Painting. It looks from the outside like just like any other shop but here I am greeted quietly and courteously at the doorway and shown into a warm, carpeted room filled with an atmosphere of calm and peace. The walls are completely covered by mysterious but beautifully coloured paintings depicting obscure Buddhist gods, and the air is fragrant with incense. Gentle oriental music is playing. This oasis of calm is solace to the fevered state I have been in for the last twelve hours, and I am pleased to become absorbed into it..

Om invites me to sit on the floor and then offers me a glass of sweet tea which I gladly accept. There is one other student here who talks quietly to Om while I sip my tea. Soon a much older man enters the room and Om introduces him as his senior painting tutor. After shaking hands, he gives me a short introduction to the philosophical import of each Thanka (Buddist painting) while the two students remove the paintings from the wall and hold them up for me to admire. The experience is not unlike being in a Turkish carpet shop and I fully expect I will be asked to buy something before too long.

After his discourse, the tutor predictably changes tack and says, "It is part of my duty to raise money for our work so that we can maintain our Buddhist traditions. So I will ask you now to buy one of our Thankas and to make a donation to this school."

I have already decided to do this, and am delighted to discover that the one I choose was painted by Om himself. It is quite small, measuring about a foot square – and the workmanship is exquisite. Om tells me that he was given some help on it by the Master. I thank him for bringing me here and say that it has been a most interesting experience - I admire him for dedicating himself to this way of life, paying his way by working as a shoeshine boy. But it is time for me to leave now, so can he show me the way? On our return into the busy world, he twists a few more rupees out of me for a bowl of rice.

"Thank you, Papa! You can't get lost now," and he skips away, grinning as cheerfully as ever.

I return to Hotel Tradition quite easily using the landmarks I spotted on the way and am pleased to be feeling more familiar with this curious network of streets at last. Back at the hotel I find Kaji waiting for me. (I do not enquire whether or not my note for Ramesh has been collected.) We board a taxi and take the drive to his home in the outlying district to the north of the city. Oh, the relief to get away from the noise and crowds of downtown Kathmandu! His is a quiet residential area with some attractive modern houses, but I can see that it is already becoming overcrowded with new developments.

Quite soon we arrive at his house and I am introduced to Kaji's wife. Amita is there as well, and she is pleased to show me her new PC on its computer table and the new swivel chair. With luck, she will have an internet connection in two weeks' time (although I later learn that it did not come for about three months).

Kaji's wife runs a little grocery store at the front of her house and this probably provides them with their only source of income. Unfortunately, Amita's sisters are not at home. This is not unusual because Sunday is a working day in Nepal, and all the schools are open. I imagine that Amita has come back especially so that she can show off the computer to me.

Kaji and I have lunch together, served by his wife: roast chicken, roast potatoes, cauliflower, prawn crackers, rice and curry sauce. When we have finished I thank her for providing such a delicious meal and then I ask Kaji if we can take a walk round his neighbourhood. We set off into the heat of the afternoon, and he points out with some regret all the new houses that have been built since he moved into the area. Even so, it is an attractive environment with large stretches of undulating countryside and many mature trees still remaining.

By mid-afternoon, I feel it is about time to depart. My lack of sleep is beginning to catch up with me, and I really don't want to start yawning in front of my hosts. So we bid each other a fond farewell. I thank Kaji for guiding me and looking after me so well, and Amita again thanks me for her computer. I take a photograph of them in front of their shop and, with a promise to keep in touch, I am soon on the local bus taking me back to the city centre. I really don't expect ever to see them again, but... how wrong one can be about the future.

I get off in the middle of the city and take a short walk through Ratna Park where a local entertainer is keeping a large circle of men amused with his banter and conjuring tricks. A nearby road is given over to an anti-Maoist demonstration and this is bringing all the traffic to a complete standstill. I gather that there are going to be further demonstrations for the rest of the week. I walk past a very grand building called the Ministry of Health and Population – what a job they have on their hands! My route takes me across the bridge over the river Bagmati which separates Kathmandu from Patan. To the east are slum dwellings and allotments, and on the other side there is a colourful gathering of women on the river bank. From here there is a steep climb to the Summit Hotel where, of course, my lovely room is all ready and waiting for me. I climb up to the rooftop garden which provides a spectacular view of the city under the late afternoon sun. I take one last picture; my camera is now full and the journey complete. I have developed a fondness for Kathmandu and feel satisfied with the way everything has worked out. But now I am not thinking about much more than a warm bath and a good night's sleep.

# Chapter Six

During the last part of 2009 after my return home, the question of Ramesh was frequently on my mind. I learnt from the first email he sent me that he never made the appointment at Hotel Tradition because the traffic in the streets of Kathmandu had been so bad that day. The way he expressed himself allayed most of my fears that I was being taken for a ride, but some disquiet still hovered in the back of my mind. Also I knew that it would be very expensive to send him to Medical College, and I was hesitant about committing myself. But on the other hand I felt a sense of responsibility towards him, and maybe our meeting was no accident. In a strange way I felt I was his lifeline, so I could not ignore or forget him, however tenuous our connection. So, I began an email exchange which maintained the link between us, and over the next few months I increasingly realised the significant role I would probably be playing in his life.

Early the following year, I received a letter from Ramesh, which contained some photos of himself and Indra Maya. He told me that he was preparing for his entrance exam to Medical College and was excited to think that one day his dream of becoming a doctor might come true.

Three months went by before I heard from him again. He had failed the entrance exam and was feeling quite devastated. But he also told me that there were some revision classes he could take, which would give him a much better chance of success at his second attempt. Could I send him some money for these classes? I replied that I would, and he gave me Indra Maya's bank details. A few days later I received an email from him, expressing both his gratitude and his determination.

I imagined that Ramesh was studying hard for the next few months, as I did not hear from him again until 30th August when he told me he would soon be taking his entrance exam. He also gave me some idea what the fees would be if he was successful.

I remember how horrified I was when I realised that subsidising the education of this unknown boy from Nepal was going to cost thousands of pounds. What was I letting myself in for? I shrank from the prospect of paying such large sums into some unknown Nepalese bank account and found myself wishing that I had never met Ramesh. My state of anxiety was in total contrast to his expression of delight in this email:

*Dear Uncle Nick,*

*Jaymashi and Namaste!*

*Today, I was very happy because I am informing you about the result of my MBBS entrance exam.*

*Due to the praying of you and your encouragement I have passed the entrance exam. I would like to thank God first of all for this. My name was there in the list of passes. For these, I would also like to thank you from my inner core of my heart.*

*Other information regarding admission, I will inform you in my next mail.*

*Once again lots hearty thank to you uncle.*

*Lots love, regard and respects,*
*Ramesh*

I replied on 12th September, expressing my sincere happiness for him, while at the same time feeling deeply uneasy about the whole affair.

There now followed a time which must have been torture for us both. I could not bring myself to send the fees for admission and first year, which in total amounted to £25,000, while he must have been anxiously waiting because the college places were filling fast following the exam results. No doubt he was worrying that he would have to wait another year if I delayed.

He even telephoned me one Sunday morning at home, and I realised then how desperate he must be feeling and that I would have to honour my pledge. So I took a deep breath and arranged for the money to be transferred to Indra Maya's account.

The transfer arrived just in time for Ramesh to be enrolled as a first year student at Lumbini Medical College. This was not his first choice, which had been Kathmandu College, but because of my procrastination all the places there had been allocated. Lumbini was an overnight coach journey away from his home and I felt quite badly about this. But I received a grateful email from him very soon afterwards:

*Dear Uncle Nick,*

*Jaymashi and Namaste!*

*Thank you very much for sending money. I received it. I have also paid for Admission and 1st year. So, tomorrow I am leaving Kathmandu for Lumbini Medical College. So, please pray for me, my journey and my studies. My class start from this Sunday, 26.sept.2010.*

*I bit busy today packing everything.*

*With love and respect,*

*Ramesh*

Once I had taken the plunge, I had absolutely no regrets about what I was doing. It was making the whole difference to this young man's life, and I instinctively felt that he was going to make a success of his training. I received the following email on 19th October.

*Dear Nick uncle,*

*Jaymashi and namaste!*

*You know uncle, you have become the special person of my life. You have occupied a special portion in my heart.*

*It was a great to be in the college. When I entered the college still I could not believe myself that I was in medical college. When I had my first lecture, then only I believed I was in a medical college and I was studying here. This was possible just due to your trust on me. You know uncle, when I had my orientation my friends had parents proudly sitting together and with them and I missed you there.*

*So, Uncle I am missing you so much and waiting eagerly for your arrival in Nepal so that you will visit where I am studying and how I am studying. Mom Indra Maya and brother Rajesh are also happy to have you in Wasta Care Centre (WCC).*

*With love and best wishes,*
*RAMESH KHADKA*

We continued exchanging emails for the next few months but it was not until May 2011 that I began to hear about Ramesh's back pain for the first time. He told me how he was receiving physiotherapy. Then in June he wrote:

*Little improvement is with physiotherapy, but still pain persists. After my exam, I will do check-up.*

Ramesh did the check-up, the result of which he explained in the following rather touching email, dated 20th June:

*As you know uncle, I have pain at the back. The pain was moving to right leg, so during my holidays, I returned to Kathmandu and did MRI. It was found the Nerve from spinal cord is compressed. Due to this compression, I am having regular pain in right leg. The treatment for this is a surgery. It costs money. So, I am afraid what to do uncle. As you know, now I am not in WCC, I am feeling really depressed. I have contacted with members of WCC but they cannot afford the cost. Now, you know uncle you are my guardian and the person whom I can trust. I lost my father when I was a boy, and I do not remember even his face. If he was here he would do something, now for me you are like my father. So, uncle, please do something. I know you have done a great thing by supporting my study but whom to ask for help now. It will be what you wish for uncle, I am not putting any pressure on you.*

*I am praying so that this pain would be reduced because everything is possible in God. I know I cannot pay you back but I will fulfil by duty towards you as a son does towards his father. I want to meet you and get a hug from you, who loves me and whom I love. I am not having a long holidays, otherwise I would come to you uncle. After my university exam, I may have 2-3 weeks holidays, so I will try to manage so that I can meet my uncle.*

I wrote back saying that I would be pleased to pay for his surgery, and I also offered to buy him a computer. I told him about my own godfather, Douglas, who had been so important in my life, and now I wanted Ramesh to regard me as his godfather. He replied the same day, saying that it was great to have someone in his life whom he could call 'Father'.

During July, Ramesh had exams so I did not hear from him again until the middle of August.

*Dear Godfather*

*Thank you so much for your letter. It was really inspiring for me because it reminded me that far away in England there is someone who always thinks and prays for me. I am so much fortunate that I have you as my godfather and this is the happiest moments I am having in my life. I thank God for being so much generous on me and showing His love through you.*

*Tomorrow, I am going to meet a doctor to talk about surgery. As soon as possible, I would like to have surgery.*

*At last, please give my lots of love and respects to aunt Adrienne. And uncle please do write whenever you feel because I am always waiting for your mails and letters. And take care and you and aunt are always in my prayers.*

*With lots of love and respects.*

*Your godson in Nepal,*

*Ramesh Khadka*

The surgery was completely successful, and after a few weeks' rest, Ramesh was fit enough to travel back to Lumbini for the beginning of the following term. But first he wrote a letter to me:

*Dear Godfather Nick*

*Jaymashi and Namaste!*

*I am really feeling great with my health after having my surgery done. I would like to thank you for being there to relieve me from my back pain.*

*I was little scared before my surgery as I had never fallen so seriously ill and I was never admitted to hospital. But I prayed to God and surgery was successful. It was possible due to your ever-loving prayers. I am so much fortunate that I have you as an important part of my life. I had never thought that someone who lives physically far but his presence I can feel every day with me. We have met once and sometimes one meeting changes our lives. Yet it changed my life and brought lots of love, affection and grace. And I believe everything is God's plan.*

*I believe we together have to do a journey. You should be there to bless me when I start my career as a doctor. You should be there to see me achieving my goal, and you should be there to celebrate every success of my life. And you know that you are my most valuable treasure too. But I know you are a happy man who enjoys every moment. This is what I learned from you.*

*Yes, you are absolutely right, dear Uncle. Nepal needs more doctors. As you know, Nepal is a developing country and you know developing country should look for development in health, education and economics. And I believe health is the most important aspect of all. So I have always dreamed of a good doctor, who alone cannot change the health status but inspires others to join hands together for the sake of country and poor people. I believe that God has given me the opportunity to become a doctor just because He has a plan in my life. He has said in Bible "You just dare to dream, I will fulfil your dream" and in my life it is coming true.*

*I have a vision to start an institute as well as a health care centre, which will provide Health Education for those who cannot afford the costs of medical studies. Moreover, I wish to serve those people who are deprived of all health facilities. I wish to go to those places where people don't see any hospitals and die due to small causes like diarrhoea. I alone cannot do anything, so I need God's grace and your prayers, Uncle. Now I haven't learned anything. I still have a lot to know, learn and see. I believe God will be there to guide me in every step, as He has done till now. And I need you to see my works for God.*

*Now I am taking rest. I am feeling so much better, so I could sit on a chair and write to you this letter. I am sorry my handwriting is not so well, but I believe we can understand each other.*

*Please give my regards to Aunt Adrienne. Please take care and do write to me. I am waiting eagerly, uncle.*

*With lots of love, respects and prayer.*
*Your godson,*
*Ramesh Khadka*

By now nearly two years had passed since our only meeting, and we both seemed to feel a strong bond between us. We were soon exchanging emails about my return visit, and I was looking forward to seeing his college as well as meeting other members of his adoptive family. But above all I wanted to become better acquainted with this remarkable young man, now my godson.

# Chapter Seven

After two flights (via Abu Dhabi), I arrive at Tribhuvan airport, Kathmandu, at 3:30 pm, as scheduled. It is March 10$^{th}$ 2012. It takes me over an hour to queue for a visa, but from then on there are no problems. I emerge into the sunlight and look around but there is no sign of Ramesh. Is he here, I wonder? Will I recognise him? Then suddenly he is rushing towards me and I know him instantly.

"Uncle!" he cries. It is such a pleasure to see him again. I drop my bags and we enjoy a big hug.

"I am really pleased you have not changed too much," I say. "I was worried I wouldn't recognise you."

He laughs a bit shyly, picks up my heavy case and we head for the taxis. It is extremely hot and I can't find my sunhat. But we are soon on our way, heading for Wasta Care Centre. We don't speak much during the journey – instead I stare out of the window at the small shops which line the street, enjoying the now familiar sounds and sights of Kathmandu. After about twenty minutes, I recognise the short hill to Wasta Care Centre which Pastor Shyam drove me down two and a half years ago. We have to stop sooner this time, as the taxi cannot make it down the uneven track. So I pay the driver, and Ramesh wheels my case along the road. It is so rough and full of potholes that in the end he has to carry it to the big iron gate which is the entrance to his home.

"Come inside, Uncle," he says, and there standing on the steps of the orphanage are Indra Maya and all the other boys waiting to greet me: Niran, Anish, Sunil, Arun and Pancha. In time I will get to know who is who, but for now I am content just to bow, say 'Namaste' and shake hands with each one. They give me a warm welcome and it is wonderful to be here at last. I remove my shoes before entering the house, and Ramesh shows me to my bedroom.

"This is for you, Uncle," says Ramesh. "Some fruit, some water, candles and matches for power cuts, towel, cupboard for clothes. Can I unpack for you, Uncle? Or do you want rest first?"

"Please could you unpack while I send a text to Adrienne to let her know I have arrived safely?"

Half an hour later we are sitting in the living room, while Mama makes me a cup of tea. The other boys do not join us and in fact never will on these occasions. It seems they prefer to stay in their room. Perhaps they are shy of me, or just don't want to intrude. I'm never quite sure.

"I have brought you all some homemade cookies. Adrienne thought you might like them."

I fetch the cookies from my case and Ramesh takes them to the boys in the other room. In a few minutes he is back.

"They say they are very good, Uncle. Please thank Aunt for her good cooking." (From now on Adrienne will be "Aunt".)

So we talk over our tea for a while and then I express a wish to go for a short walk after sitting for so long in an aeroplane. Ramesh leads me on an introductory tour of the neighbourhood which comprises sandy tracks, a dried-up riverbed, a barren bit of undeveloped land, some simple dwellings, a row of open-fronted shops and a large building called Dr Livingstone Secondary School. We talk continuously, sometimes stopping to reflect, or take in a view, or expand on a particular point. Conversation is obviously going to be no problem.

We try to catch up on everything that has happened since we last met. I also learn a little more about his childhood, and his adoptive family, which comprises Indra Maya's five sons

and one daughter. I discover that he is 22 and his birthday is on 13th February – so he is almost a Valentine boy and therefore the subject of much teasing. He is easy company and I feel we are getting off to a good start.

We return home where supper awaits us - *dal bhat* (pulses and rice) with mushroom curry. This has been cooked by Niran who I learn is the best cook in the family. Niran has a big round face with a jolly smile – he would make an excellent Father Christmas – and he stands in the doorway hopping from foot to foot, beaming at us.

"Let us pray before the meal. Uncle, would you like to say a prayer?" asks Ramesh.

This is not a custom I am used to, but I do my best: "Heavenly Father, thank you for bringing me safely to Kathmandu today to see my godson Ramesh. Bless these ten days we will have together. And bless this food, and may we use it in Your service. Amen."

"Amen."

The meal is excellent and lasts a long time because we still seem to have a lot to say. This will be the case at all our meals together and in time I get used to eating food that is considerably less than hot.

Afterwards, I offer to take the plates out and wash up, but my offer is refused. In fact, the kitchen will remain totally out of bounds for me during this trip. Instead, I bring out a folder of old family photos and these are of great interest to Ramesh. He wants to learn every detail of all my family members, and writes their names down carefully on the reverse side before continuing to ask me for more information.

"I will be looking at these photos very often and remembering their names and faces," he says with a smile.

By now it is nearly 9:00 pm and I can hardly keep my eyes open. Ramesh notices this.

"Now you must rest, Uncle, after your long day. We will carry on with more photos tomorrow."

He is so solicitous. He all but tucks me into bed when I am ready. "Good night, Uncle, God bless and sweet dreams!"

Sleep comes very easily tonight.

We have planned to set off for Lumbini Medical College the next day. One of Indra Maya's sons, Karjun, runs a taxi business and he will drive us there. The journey will take at least ten hours, so we will probably not make it in one day. And because I want to visit the birthplace of the Buddha at Lumbini (now a World Heritage Site), we will probably be away for three nights.

I need to change my dollars into rupees. While Ramesh and I are discussing this and drinking grape juice outside the front door, their landlord, who lives on the first floor, appears on the scene. He wheels out his shiny red Honda 1000 cc motorbike, which I duly admire.

"Let me give you a ride on it," he says. "I am just setting off for work."

"Can you take me to the bank?" I ask.

"Of course. Climb aboard."

He will take me to Jawalakhel where there are plenty of banks and Ramesh will run after us. This seems a bit unfair, but Ramesh is quite happy about it.

"I am a fast runner, Uncle. I will soon be there with you. Just wait by the market."

So I find myself again on the back of a motorbike and enjoy feeling part of the daily life of Kathmandu, as we join other people making their way to work. Eventually he drops me off and shows me where to wait. This is obviously a popular meeting place and there is plenty to interest me as I stand at the busy junction. There is a big outdoor market here, and as I look around I see someone apparently giving ear treatment to a friend. Does he have earache, I wonder, or just blocked ears? Ramesh appears and takes me to the bank, where I change my dollars into rupees. Soon we are jogging back home for an early lunch because we need to get going by 1:00 pm. It is all very exciting. I think Ramesh is as keen to show me his college as I am to see it.

Karjun is with us by 12:30 and I am introduced to him and his red saloon car, in which he takes much pride. The car must be at least ten years old for it has a somewhat battered look and the paintwork is faded, while the passenger door no longer opens from the outside. I understand that he is not getting much taxi work at the moment, so this stretch of three days' employment will earn him some welcome income.

"Good to see you, Nick," he says with great delight. "Thank you very much for asking me to take you to Lumbini. I promise I will give you excellent service and bring you back safely."

"I am looking forward to it."

He stows away my bag (I notice that while I am taking a sizeable bag he and Ramesh seem to be taking almost nothing) and after saying goodbye to Mama and Niran, we set off.

Roads and road users in Nepal do not have a particularly good reputation, so it is a fairly hazardous journey, during which I opt to sit in the back seat trying to enjoy the scenery rather than take notice of the many crazy lorry drivers, the cliff landslides or the fearsome potholes. There are plenty of impressive hills to admire - and soon a wide flowing river, profuse vegetation and the occasional cow straying into the road provide further distraction. After three hours I am glad when we pull in to the Riverside Spring Resort Gardens Hotel, which proves to be an oasis of calm and rest – "an excellent place for a honeymoon" I tell Ramesh, but this comment appears to embarrass him somewhat. We all have a cup of *chai* (because it is 4 pm, at which time my English habits assert themselves) and despite a headache from the heat, I begin to feel a little better.

There are several more hours' driving ahead before we pull in at the Stadoor Hotel at Budiral. They have a single room for me and a twin for Ramesh and Karjun. After showering, we meet in the dining room for a late supper. The menu is all in Nepalese, so Ramesh tells me what I can and can't eat (he even tastes the food for me when it arrives). He is determined to take good care of me, and next morning when I wake up my headache has gone.

This day fulfils the main purpose of my trip to Nepal, because by 10 o'clock we have arrived at Lumbini Medical College. First of all I am taken to meet some of Ramesh's tutors and professors. They all make polite enquiries about where I come from, and one of them even congratulates me on being Ramesh's sponsor, while others speak highly of his progress. Ramesh shows me round the campus and I soon realise this is a college in the early years of its development. While there are several multi-storey buildings comprising laboratories, teaching rooms and student hostels, there remain large areas of land which are still being developed. It gives me a feeling of both pride and pleasure to be walking round with him, knowing that his dream has come true and that my decision to sponsor him has paid off.

This is clearly a thriving establishment. I meet the college Principal, Professor Vikram Singh Chowhan, who comes across as a man of great vision and enterprise. He tells me that the college has its own hospital where Ramesh will gain invaluable training. There is a large staff of mainly Indian doctors who are teaching here in their retirement, and they are both experienced and committed to the education of the students. All this in a healthy unpolluted environment, which will do Ramesh so much good.

We walk over to the students' canteen area where we find some of his friends. They are all very happy to see him and he introduces them to me. He seems so at home here and appears completely relaxed and at ease. As we walk away, I ask him a question which I hadn't thought of until now:

"Are you missing some of your lessons to be with me?"

"Yes, Uncle. I asked for special permission to be with you for these ten days while you are in Nepal."

I am slightly amazed and touched by this. "I hadn't realised..."

"Don't worry, Uncle. I will easily catch up. It was more important to be with you. I had to be with my godfather." He smiles. "Come, I want to show you the hospital where I will be doing my rounds next year."

# Chapter Eight

By the end of the morning we are back on the road, on our way to the nearby town of Tansen and our hotel, The White Lake. Tansen, which is situated on the side of a hill above the Medical College, is both picturesque and clean (I note plenty of small blue dustbins dotted around).

Our hotel rooms have balconies with fine views across the valley. Karjun wants to rest, so Ramesh and I walk out into the town. I have a craving for an apple, so we buy two from a street vendor. She washes them for us in a small black bag. We munch them as we wander round the town – being set on a hill the streets are a steep climb and we are soon in need of a drink. We come to a square where, in one corner, we notice a restaurant called Nanglo West with an inviting courtyard set out with table and chairs. There are bougainvilleas clinging to the walls, and several colourful shrubs in pots which lend the place an attractive restful atmosphere. We sit down here and are soon served the delicious sweet milky tea which I love so much. The place also has wi-fi, so Ramesh speeds back to the hotel for his laptop so that I can send Adrienne an email. While I do this, he makes corrections to my diary and fills in the names of his brothers (as he calls his fellow orphans), as well as the names of Indra Maya's family. This will be invaluable for me in due course.

We wander round the town for a while and come upon a football match which we gather is being played between the army and the police. At half time, the score is one all – but only because the referee awarded a rather dubious penalty in favour of the police. Ramesh clearly loves football. He was captain of his school's 1st XI, but cannot play so well now since his operation. He is an ardent Liverpool supporter. Strange that English football should have such a global following.

We leave before the end of the match and climb the hill above the town until we reach a small Hindu shrine. There are good views from here, but it is getting dark and cold now, so we return to the hotel, collect Karjun and go back to the same restaurant for supper. It is a very old wooden building, and inside it has minimal lighting, lending it a quiet brooding atmosphere. Over supper we spend the evening talking about our backgrounds and childhood days. I learn that Karjun's father (Indra Maya's husband) died when he was eleven, so being the eldest he had to prepare all the meals for his five brothers and sister while Mama went out to work, selling fruit and vegetables at a tiny roadside store. Ramesh tells us how, as a young boy, he would sit up very late to finish his homework, refusing to eat or drink until it was all done properly. This made him a great source of worry to his family. I tell them a little about my own background in England, which could hardly be more different from theirs, and we reflect on the strange circumstances that have brought our lives together. It is all very mysterious, but wonderful as well.

We walk home and Tansen is now dark and quiet – there is not even a dog barking. Ramesh will listen to songs on his phone to send him to sleep, while I have my diary to write up.

For some reason, though, I do not sleep for long – I am up well before dawn, walking the streets. A few cockerels are crowing and there is a hedgeful of sparrows twittering. By seven o'clock the sun is up and the valley below is starting to shimmer in the early morning light. At breakfast, Ramesh asks me about my trip last year with Adrienne to Greece. I tell him about Asclepius the healer and the healing sanctuary of Epidaurus, which he knows something about from his lectures on the history of medicine. Then I start telling him about the oracle at Delphi. A young Dutchman at the table next to ours pricks up his ears at the mention of prophecies.

"That is my job too, you know – making prophecies. I run a software business, designing programmes to predict future stock prices."

"Is it successful?" I ask him.

"Quite successful, yes, thank you. That is why I am able to take a month off trekking in Nepal."

"What is your prediction for our future today?"

"Hmm, let me see... I predict you will get very wet before the day has ended."

I express my disbelief when the day is looking so fine and we part company. Soon we are in the car driving to the birthplace of the Buddha at Lumbini. We check in to the Buddha Bhoomi Hotel and hire bikes for our ride around the World Heritage Site. It is a warm sunny morning and we cycle down the driveway lined with blossoming trees. There is a lake nearby and the atmosphere is calm and still. Prayer flags are strung up between trees and monuments, fluttering gently over sacred pools of water. We walk around the holy site and enter the Maya Devi temple (Maya Devi being mother of the Buddha), where we see the very stone on which Gautama is said to have been delivered.

There are still many temples to see but we need sustenance, so we return to our hotel for a lunch of vegetable noodles and a glass of lemonade. The sky looks a bit threatening now and soon we see drops of rain.

"What do you think, Uncle? Best to wait, I think, in case of a downpour."

"No, no, I am sure it won't amount to anything. Let's risk it."

This is a big mistake. Only a few minutes after setting off on our bikes, the rain comes down in torrents and we are soon soaked through. We now find ourselves spending the afternoon visiting Buddhist temples and stupas in a totally drenched state. Fortunately we see the funny side, particularly when we have to remove our shoes and paddle across temple courtyards. Eventually the rain eases and the clouds part to reveal some sunshine. It isn't cold so we manage to dry off a little. The site covers a vast acreage, and there are modern temples here built by Buddhist foundations from all over the world. Some are completed and decorated in bright colours from roof top to their lowest steps, while others have only just been built and are in a totally unadorned state of hard greyish wood, waiting for the paintwork to be done. Either way, they are uniformly large and impressive.

By 4:00 pm we are beginning to feel cold in our damp clinging clothes, so we head back to our hotel for a shower, change of clothes and a long rest. It is then that I remember our Dutch companion at breakfast this morning and his prediction…he was right after all!

The next day we set off on the long journey back to Kathmandu. This has been an exciting time for Ramesh. He has never been away on holiday before, staying in hotels and being served meals of his own choice. He has shown a real delight in an experience which most of us in the West take for granted.

Karjun has two interesting detours for us to break the journey. The first is to the remote rural village of Bhawanipur, believed to be the birthplace of Maya Devi. The site of the palace where she was reputedly born is on the edge of the village and much of the excavation work is grown over. But there is a small 19th Century shrine containing plaster figures of Maya Devi sitting beside seven of her sisters. Surrounding the shrine is a grove of trees joined together by prayer flags. Karjun talks to a local guide while Ramesh and I amble round the site, enjoying its tranquillity. The warmth of the morning sunshine is finally drying off our clothes, still a bit damp from yesterday.

The next stop is a vast Bodhi (or Sacred Fig) tree growing in the middle of a wheat field. It has a girth of at least thirty feet and is equally high. Under a similar tree at Bodh Gaya in India it is believed that the Buddha sat until he achieved enlightenment. A local farmer explains to Karjun that from afar this tree looks like a great hill. After a leisurely tour of inspection we drive off and then stop briefly to look back – and indeed, it does look like a hill!

We are soon back on the road for another long stint. The next event is a flat tyre, which Karjun deals with neatly and quickly by swapping wheels. But now we must get the flat tyre fixed, because to be without a spare on the long remote ascent to the rim of the Kathmandu Valley could be disastrous. Fortunately, he spots a tyre shop in a village called Timara, and we watch with interest while the men set about changing the inner tube. I pay for the repairs, and we safely resume our journey. Soon, at a petrol filling station, we have a break for peanut cookies and a fruit drink because the home stretch is about to begin.

After driving along the riverside road, we begin the long, twisty and somewhat hair-raising ascent. Cars and lorries overtake each other alarmingly on blind corners, and as darkness descends, all common sense seems to desert the road users, desperate to get home quickly. I keep my eyes closed most of the time trusting that Karjun is a safe driver. But with a headache coming on, I long for this final leg of the journey to end. It is a slow, bumper-to-bumper drive, and we eventually arrive home at eight o'clock. Mama and the boys are waiting for us at the front door, delighted to have us back, and there is a welcome supper waiting for us, prepared by chief cook Niran. It has been a great trip, but I am pretty tired and very thankful to be back home.

✸ ✸ ✸ ✸ ✸

# Chapter Nine

The long drive to Lumbini and back has given Ramesh and me the chance to get to know each other better. I remind him of the time we first met and he told me something of his early life. Can he continue the story for me? I want to know how he came to be taken in to Wasta Care Centre.

"I will tell you the truth, Uncle, because it is important you know what happened, even though it is a sad story really. After the death of my step-father, we all came back to our village. Then I worked in the field with my mother to feed two brothers and one sister. They were very small and I was only elder then. I used to plough the fields and sometimes I could not handle the oxen as I was small for that work. Once, when I was preparing shed for cattle and cows, I poked my forehead with a metallic lever. I fainted and when I woke up my mom was there in front of me and she tore her sari and tied my forehead. And I did not go to hospital or a health post as we did not have money for that and nobody was there to help us. When I remember this incident, I still feel to cry. And I have still that scar on my forehead and I can feel the dent there. I was really hard-working boy but I was only paid half of the wage that adults used to be paid. Then one day my uncle, my maternal uncle who stayed in Kathmandu and had taken my elder brother with him and put him in one of the orphanages, visited our village and told me that he will search for orphanage house for me as well. But I wanted to be with my mother and help her. You know, in the village we did not have house to stay in. We stayed in my maternal uncle's house but later they asked us to go somewhere else. And we moved to another villager's house. They gave us shelter for sometime but they started disliking us. We did not have a single meal without my mother crying. I was smaller and they used to treat me badly. Some people who liked us gave us sometime food to eat. My brothers and sister used to cry for food. Mother used to work with me and bring flour with her in the night and make porridge. We did not have curry and we used to eat with curry of dried leaves of spinach. Those days were so worrying and breathtaking for me. Whenever I remember those moments, my body still shakes in fear. When it was about six months time later, I was told to go to Kathmandu by that maternal uncle. Then, I did not want to because my mom would be alone and whole villagers would dislike her and there would be no one to support her. But my mom is a strong woman and told me that she had struggled throughout her life and it's her fate, but she did not want her sons and daughter to live like her. So she strongly said I would not be staying and working in village with her. She would work and feed my brothers and sister but I should be going to Kathmandu and get chance to study. She knew me very well and she believed in me that if I could get chance to study I would do good. So, she sent me to Kathmandu. You know, in my village there is still the system of untouchables. When I was there, I had asked food from them and had eaten. I did not have any new clothes to put on and no slippers. I came to Kathmandu in the same clothes which I had worn at working in field. I still remember I put on the slipper which had a hole at the sole and it was my mother's. You know, I did not have money for the ticket. I stood in the bus till I reach Kathmandu. Those days of scarcity and poverty I will never forget even though I try. I feel now that those days were a lesson for me to learn and I have learned how it feels to be poor and orphan. I never wish anybody to be poor or orphan because it squeezes my heart and fills my brain with madness."

He first came to Wasta Care Centre at the age of nine, and the care and attention he received was such that he never wanted to leave the home. Whenever an outing was arranged, he preferred to stay indoors and watch TV. He must have felt so unloved as an infant that now he badly needed to make up for the deprivation. And when his school days started, he made the very most of his opportunities and found every subject equally fascinating. It seems he was a model pupil, coming top in all subjects and having an exceptionally retentive memory. He also shone in sport and eventually became captain of both the cricket and football teams. As the second oldest and brightest of the five orphans, he feels a great sense of responsibility to his brothers and is very protective towards them. Although he is naturally shy and lacking in confidence, he has a leadership quality which is apparent to others – and at Medical College he was soon proposed as class representative for his year group.

"I really don't want to be," he said. "Maybe not," his friends said, "but we want you to be," because they could see that he would be fair and responsible. He told me that he always tries to act according to the dictates of his conscience and feels that those early years of deprivation and hardship were given to him as a reminder of the suffering other people endure. One day he would like to do something for his country and becoming a doctor will certainly give him the opportunity to do that.

By contrast, my own childhood and early adulthood of privilege and good fortune did not leave me with any sense of social responsibility or a strong desire to improve the lot of others. But I do feel I can share some of my interests and passions with Ramesh, and occasionally I find myself launching into a brief introduction to the life and works of some of my favourite writers and literary figures such as William Shakespeare, Tolstoy and Dr Johnson. He seems to be interested in everything I tell him about England, English history and European culture and I keep saying, "When you come to visit me, I will take you to this place and that place, and you will come with me to hear a concert and I will show you my favourite piece of coastline" (he has of course never seen the sea). So we do seem to be getting along fine, despite my early apprehensions.

Now I have a week to spend at the orphanage. Because Ramesh sleeps a bit late sometimes, I find myself going on early morning walks with Niran. His English is not as good but I like his outgoing personality. He is very friendly and likes these walks with me because he has to spend so much time studying at home by himself in preparation for his forthcoming exams. He is training to be an accountant and is in his second year. However, all the students at his college are expected to spend years three to five studying in Delhi and he is worried about the finances. Also, he will have to buy more books and his computer is very out of date. I feel that as I have helped Ramesh, it is only fair that I help Niran as well. He remembers the time I came to Golgotha Church and told the story, 'Two Frogs in Trouble'.

"I like empowering stories like this one. When I teach the little children at Sunday School, I like to tell them stories like this. Can you please send me some, Uncle?" I say that I will.

Today I have asked Ramesh to take me to Philip Holmes' Mosaic Workshop. I learnt about this workshop through my own interest in mosaics. The story is that, following the tragic suicide of his wife Esther Benjamin, Philip retired from his job as an army dental surgeon to establish a charity in her name. He started to learn about the plight of many Nepalese children sold into slavery and forced to work in Indian circuses. Some of these children were orphans and several were also profoundly deaf. So, with a team of volunteers, Philip would go to the circuses and rescue these children, bringing them back to Kathmandu where he was living. The next task was to give them some means of making a living, preferably something artistic and creative. He realised that there was no tradition of mosaic art in Nepal, so he returned to England and went on a mosaic-making course. Once back in Kathmandu he was able to teach this craft to the children, give them a safe haven and provide them with a livelihood.

Ramesh makes some enquiries and takes me to the Esther Benjamin Trust Office where we meet Abha who works for the charity. She is going to take us by taxi to the village of Godavari to see the workshop. Imagine my delight when our taxi driver turns out to be Karjun. This family has a strong network of communication by mobile phone, I am beginning to realise. They all know what they are all doing and so are able to help each other whenever necessary. I am constantly touched by their family bond of love and support.

Godavari is a 45-minute drive away, set in some attractive countryside on the edge of the city and popular with picnickers. We eventually arrive at the mosaic workshop, a very simple one-storey brick building located in a remote area up a dusty track, high enough to enjoy a fine view over the distant city. It is a hot, sunny morning. Two of the craftsmen, young men in their mid-twenties, are working on the terrace. Abha introduces them to us, and we greet them with 'Namaste', but all vocal communication ends there, as they are completely deaf. Instead, their mosaics speak for them: so detailed, colourful and refined. They are working according to the designs of other artists, depicting mainly animal and bird images. Inside the building we see a small gallery of finished work and are introduced to a group of girls cutting up tiles.

"Once they have had some instruction, these girls are wonderfully dextrous and skilful," Abha tells us. "They can cut out a neat circle in a minute with these tile cutters – and it will always fit perfectly into the pattern."

"What materials do they use?" I ask.

Abha leads us to another room where in the corner lies a vast heap of broken bathroom tiles.

"This is it?" I ask.

"That's all they use. We are given these cast-off tiles by bathroom shops and they serve our purpose very well. They are easy to cut into shape and are completely free."

"And are they able to make a living making these mosaics?"

"One mosaic may take four weeks to make from beginning to end and they are paid piecemeal. With their earnings they can pay their rent and make a small living. They are grateful for the work because it is such an improvement on their previous lives."

This is an impressive achievement by Philip Holmes, who has turned the tragedy of his wife's death into such a worthwhile charity. I learn that he will be running his second London marathon this year to raise money for the Trust. Hopefully, I will get the chance to meet him one day.

Ramesh is equally impressed. "I never knew this place existed. You are showing and telling me so many new things, Uncle."

Nearby is the Kathmandu Botanical Gardens, and as none of us has been there before, we decide to take a stroll round. There are no labels on the plants so we amuse ourselves trying to guess at their names. Abha and Ramesh chatter away easily, as they have done in the car, seeming to get on well. I learn later that she is a dropout medical student, so they have some experiences to share.

After this visit, Abha wants us to see the print and jewellery workshops of the Esther Benjamin Trust. Karjun drives us to a newly-built artistic centre in the city, where we are introduced to more artists working for the charity. I admire the silver jewellery in particular. These products are selling better than the prints which they may have to give up, Abha tells us. Soon we take her back to her office and, thanking her for an interesting morning, we return home for lunch and a rest.

I have decided I would like to buy Niran a new laptop, so at about 3:00 pm, Ramesh takes me to the commercial area of the city where, three years ago, Kaji's daughter bought her computer. After protracted haggling and bargaining to absolutely no effect, we end up buying a shiny bright blue Dell computer, which we think will please Niran. The bus ride home from Ratna Park is incredibly crowded and we are squashed into seats at the back. Ramesh tells me about his love for maths and science. It reminds me of a TV series I once saw called *Orbit* which was all about the solar system and the effects on the earth's surface of gravity, tides, oceans and winds. He is fascinated by this and I tell him I will send him a copy on DVD.

That evening, over supper, I tell him what I have been reading in that day's edition of *The Himalayan*, a short English newspaper Ramesh buys for me each day. There is always a 'health of the nation' article, which today is about positive thinking. Then there is the Prime Minister's pledge for hydroelectric power (a perennial topic which would surely be the answer to Nepal's future if only the political will were there), as well as news of the first hat-trick in an international cricket match by the Nepalese bowler, Shakyantok. We talk and talk, while as usual our supper gets cold on our plates.

Later we go through more of my photos and again Ramesh annotates these with scrupulous care on the back. I tell him a few English jokes, which take some explaining, and he reads to me from his school English Literature text book – I think we could go on all night but Mama tells us to go to bed. She is very sweet and smiling about this, but I now realise we are in the room where she will be sleeping. She makes up a bed on the table where we have been having supper, because in fact I have usurped her bedroom. I feel slightly awkward and embarrassed about this, so I mention it to Ramesh later, but he only shrugs his shoulders.

"Don't worry, Mama is fine in here. She is very happy, Uncle."

I am to spend the following day with Rajesh (pronounced 'Rajeesh', not to be confused with Ramesh), Indra Maya's third son. He is 33, married to Soni, and they have two adopted children: Sara (12) and Basanta (10). Rajesh follows in the family's strong Christian tradition and is a pastor to a rural parish, north of Kathmandu, called Sangla. He has a strong personality and was responsible for the education of the orphans in their early days at Wasta Care Centre when they were boys of 3, 6, 9 and 10. I suspect he was a bit of a taskmaster, teaching them with a firm hand. He is a short, tough-looking man with a moon-shaped face, a broad smile, twinkling eyes, and a great laugh. I take to his warm, all-

embracing personality, and I am sure we will have a great day out in his home and his parish, giving Ramesh a break from my company.

"Good morning, brother Nick," he exclaims as he climbs off his motorbike in the front yard. "How are they looking after you here? Are you eating well? Sleeping well?" and he gives me a great bear hug in his leather biking outfit.

"I am fine, and having a lovely time. I am being looked after very well."

After farewells from Mama, Ramesh and Niran, (I never seem to see the other three boys, because they have to go to school so early – and when they come back they disappear into their room) we shoot off into the crowded streets of Kathmandu, heading north. Twenty-five minutes later, we are leaving the city behind and entering a more rural part, close to where Kaji, my original guide, lives. Soon we are climbing into a hilly area and entering the rural parish of Sangla. This is the most attractive part of the Kathmandu Valley I have visited so far: undulating hills with flat plains between, farm buildings dotted around, green lush fields, arable crops, and, very importantly, fresh clear air.

We visit his small church which comprises the tiny ground floor room of a house with a metal up-and-over door. In effect it is a converted garage.

"Sit down, Nick, and I will sing you some of our songs of praise."

Singing in a strong tenor voice, he accompanies himself on the guitar in three hymns which he clearly loves. When he finishes, he puts down his guitar and says:

"We will all be having a worshipful time together, Nick, before you go, and I will be teaching you these songs."

We set off from here on a day of country walking to visit his parishioners in their homes.

"This is my work, to visit and talk to them about the Christian message and have worship with them. They are simple people who have been brought up as Hindus and still have many superstitious ideas. It is a struggle for me to convert them to Christianity, because they are so weak in their faith, but I urge them to come to church and perhaps five or six come to worship on Saturdays."

I ask Rajesh how he became a Christian.

"When I was a little boy, I was always very sick with many illnesses and finally came so close to death that Mama Indra Maya bought wood for my cremation. This was when I was 13. She gave me a Bible to read and told me about praying, so I decided to pray for myself. I did not really know how to pray but I placed the Bible on my head and prayed as best I could. Instantly, I was better! All my ailments disappeared and have never returned. So that is how I became a Christian."

All this time we are walking along paths over the fields in the morning sunshine and for me this is the most wonderful day. I am warming to this young man, so fervent and open-hearted; he has a strong personality and a very direct way of speaking.

"Later on," he continues, "I wanted to marry a Hindu girl. We were very much in love, but her mother was angry. 'You are a Christian,' she said, and refused to let us marry. So very gradually I introduced Soni to Christianity, and then patiently and slowly I talked to Soni's mother, and she gradually came to understand and appreciate it. Eventually, she allowed me to marry her daughter.

"Now I had to work on Soni's seven brothers who were also Hindus, and this was a very difficult struggle for me. But one by one I spoke to them slowly and patiently about my Christian beliefs, just like I did with their mother, and they finally changed from being angry with me to being accepting. And now they all very much love and respect me."

We call on three separate families on our parish rounds, each one living in a simple brick house surrounded by countryside which they farm. After a preliminary chat, we sit in a small group and Rajesh leads some prayers and we have a song. I am also invited to say something. Fortunately, I have one or two stories prepared for my next visit to Golgotha church on Saturday and Rajesh translates these for me. At one of the homes there are three children; originally there were five here but this was too many for the parents to support, so Soni and Rajesh, having no children of their own, adopted two of them. Naturally, from that time on, there has been a particularly strong bond between the two families.

After our visits, we stop on a hillside for a picnic lunch (Rajesh has bought some pastries from a bakery for us) before walking further up the road into the mountains to catch the magnificent views across the valley towards Kathmandu. There are several gangs of road builders up here, as this road is a totally new construction. Wives and children are in evidence as well, the women providing food and the children doing small jobs like breaking up small stones or carrying light loads.

I have offered to buy Rajesh a new motorbike, as his present one is becoming quite unreliable nowadays, and he needs transport every day for his work. First we go to Thamel where I have been commissioned to buy Nepalese pashminas and socks made of Yak wool for my English friend, Barbara Datson, who runs the charity 'CHANCE for Nepal'. She will sell these at home for a good price and make some money for her charity. We accomplish this successfully (Rajesh drives a hard bargain!) and then we go to a motorbike shop to inquire about the new Yamaha he wants. It is nearly closing time so he will go back on Sunday to complete the part exchange.

✸ ✸ ✸ ✸ ✸

# Chapter Ten

That evening I am given a special supper: curried chicken in a piquant sauce with mixed vegetables including asparagus, carrot, cauliflower and potatoes. While I eat this, Ramesh tells me more about his school days.

"I was really so shy and never said anything, but I always sat at the front of the class, because I was interested in everything the teachers taught me. I hated the pupils who disrupted the class and did not work, though this made it a bit difficult for me sometimes. My teachers saw that I was bright and really encouraged me. I was always top of the class because I was naturally hard-working and never stopped until I could do something perfectly. I am also fortunate to have a good memory: I can read a page and remember it all – even the commas and full stops."

Ramesh tells me he is trying to become more extrovert by opening up to people, talking more confidently and showing an interest in them.

"I am interested in everything you are telling me, Uncle, and I am learning so many new things. If and when I have the time, I would like to go to the mosaic school for the deaf and learn their craft; also I would like to go to Triple Gem School and do some maths teaching and keep an eye on Manish for you. I remember being told years ago that the brain is like elastic – capable of being stretched in all directions.

"But now I want to read you a short story. It is by Chekhov – we read it at school and I really liked it. So get ready for bed, Uncle, and I will come and sit on the bed beside you and read and then you can go to sleep."

This is a treat, to have my godson beside me reading his favourite short story in his carefully articulated voice. Typically, it is both funny and sad, like so much of Chekhov, and it is all about unfulfilled love. He reads it slowly and carefully, enjoying his ability to speak in English as well as conveying the story's variety of emotions. It is a delightful moment.

The next day, I put on my best clothes, the boys shine their shoes and Mama puts on a sari, because we are going to Golgotha Church. The service seems to be much as before, the only difference being that there is a choir comprising three lovely young ladies, beautifully dressed in identical saris, leading the singing. Now I am beginning to recognise some of Indra Maya's family: there is her fourth son Amosh, another pastor, helping to lead this service; and I can see Radheshyam playing the guitar and singing his heart out up on the platform.

Karjun is in the congregation with his two boys, Ashish and Abhishek. I will be introduced to them later, after the service. As before, I am invited to speak to the congregation.

I tell the story of the famous Italian Renaissance sculptor, Donatello, who was offered a large piece of marble but rejected it because it contained a flaw. "Offer it to that chap Michelangelo; he won't notice the difference." Michelangelo accepted the marble, from which he carved his magnificent statue of the young David. The moral of the story is that despite flaws and weaknesses we humans also have the capacity for greatness within us.

Walking home after the service, Ramesh asks me more about Donatello and Michelangelo, as he had not heard of them. So I tell him what I know. I also talk about Leonardo da Vinci and the great rivalry between him and Michelangelo.

"I know a story about Leonardo, Uncle. The man who posed as Christ for Leonardo's painting, *The Last Supper,* was later used as the subject for Judas. His life had gone downhill so much. Which goes to show how careful we must be in this life, not to slide downhill."

When we arrive home and open the gate into the courtyard, the fourth boy, Sunil – whom I have hardly seen – is there doing my laundry in a washing-up bowl, squeezing out my socks and shirts, hanging them on the line to dry. I hardly know what to say, except 'Thank you.'

"Thank you, Uncle," he replies with a smile. "I am very happy to be doing this for you," and I think he means it.

There is a blackout as usual in the early part of the evening. I am in my room and light a candle. There is a knock at my door. Mama brings me a huge bunch of black grapes and about ten bananas!

"Would you like some tea?" she asks me.

"No, thank you."

Despite that, in five minutes' time, tea arrives as do Sunil and Ramesh. Sunil looks a bit sheepish as he has just had rather a savage haircut.

"Where did you get that done?" I ask him.

"Just up the lane we have a barber shop."

I tell them the story of my haircut the time before and, like the pupils at Triple Gem, they find it hilarious.

"You overpaid him so much, Uncle," says Ramesh rather reprimandingly.

"Well, what do you pay, Sunil?"

"It costs me 50 rupees and that includes a head massage."

"That's about 40 pence! Please take me to this hairdresser tomorrow and seat me on his chair."

Ramesh promises to do this. I know it will be another great haircutting experience.

Saturday and Sunday nights are football nights. Whatever else this family may not possess, they do have Sky TV – so here I find myself on a Saturday evening in Kathmandu watching Swansea play Fulham live from the UK, in the second round of the FA Cup. We are all totally glued to it for 90 minutes. Fulham are outplayed, losing 3-nil on their home ground – a bit of a humiliation.

I enjoy an early morning walk with Niran the following day, and he tells me an entertaining story about a master who had two servants. The master says to them:

"I want you to carry some rocks for me from the bottom of this hill."

The first servant brings up sensible-sized rocks, but the other, trying to be clever, only brings up tiny rocks so that he will not tire himself out.

"Here is your reward," says the master, and gives sensible-sized pieces of bread to the first servant and tiny pieces of bread to the other.

The next day, the master makes the same request. The first servant responds as before, with average rocks, but the second servant just brings up one huge rock.

"Well done!" said the master. "You will both receive as much land as the distance you can throw your biggest rock," and again the second servant kicks himself for being so foolish. He can only throw his rock one metre!

When the master makes the same request on the third day, the second servant again thinks that trickery will serve his purpose best. This time he carries up one small rock and one large rock.

"Well done again!" says the master. "This time I will give you shoes which match the size of the rocks you have brought me. I hope they fit!"

Niran is quite a joker. He shows me the prison near their home which some political prisoners tunnelled their way out of a few years ago – not exactly a top security establishment! And he tells me about the time he told a friend of his that he lived in this area (which is called Nakhu), and the friend said, "Oh, so you live in a prison, do you?"

Today I am going to take Ramesh to a church service for the local English-speaking population of Kathmandu. It is the Kathmandu International Christian Church, where we are greeted by a lady called Anne Penn, whom I met here on my first trip to Nepal. Although the service is clearly western in style, it does have some of the exuberant atmosphere of yesterday's Nepalese service. The new American minister is particularly good with the children. He asks them who are the characters in the Parable of the Good Samaritan and he is given all the right answers, including, of course, the donkey.

"But now, because today is Mothers' Day, we must remember the mothers of these characters. As you know, every single person in the world, including the people in the parable, has a mother."

One bright spark pipes up, "What about Adam and Eve?"

Loud laughter from the congregation.

"There's always one!" comes the minister's wry response.

Halfway through the service, newcomers are asked to introduce themselves, so I stand up and say, "I am Nick Morrice from England. I am here to see my godson, Ramesh Khadka, whom I am sponsoring through medical college."

There are a few claps, and I feel very proud to be here with Ramesh.

The minister's sermon is excellent. He talks about the 'inconvenience of love', meaning that the need to show love to another person often comes at a difficult time, but that is when we need to show it the most. He has an easy, informal manner which is very engaging. Ramesh tells me after the service how much he enjoyed it, adding that he will try to go there again in future.

I feel certain that Ramesh would appreciate having a camera so I offer to buy him one. In the afternoon he takes me to a shopping area where I can catch up on emails and have a Google chat with Adrienne, while he checks out the camera shops. He comes back to the internet café quite excited.

"I have found one, Uncle. I hope it is not too expensive."

"I am sure it will be fine. But first I would like you to send an email to Adrienne."

He rushes off a quick message about what a great time we are having and how he is looking after me as well as he possibly can. The camera he has chosen seems a good choice, and I think he is delighted. We pay a visit to Pastor Shyam Nepali at Amit's home, where we chat and have tea. On our way home I remind Ramesh about my date to have a haircut. He takes me to their local barber shop, where a number of men have gathered to enjoy a lively debate about something or other. (I guess it is about politics, but he tells me later all that they have been talking about is their respective body weights!) However, the discussion does not deter the barber from doing a first-class job - under Ramesh's close supervision, of course. There is gentle hair-pulling as well as hair-cutting, a head massage, eyelid massage, nose massage and, finally, detailed razor-shaving around my ears and neck. I pay 100 rupees, which is twice what I should have done, so the barber is well pleased.

Back home, the shampoo has been put out, a bucket of warm water prepared, and I have a wonderful hair wash with Ramesh ladling the water over my soapy scalp. I feel totally spoilt and pampered, especially since I know that he washes his hair outside in the courtyard under a cold tap.

That evening we all gather in the living room to sing hymns and choruses. Mama chooses what we sing, Sunil plays the guitar while Arun and Anish are on drums and tambourine. I would never do this at home, but it feels just right here. I feel comfortable and welcome in their company, particularly as I am getting to know Ramesh and Niran quite well. The other three are still shy of me, just as I am of them. But maybe this will change one day.

Tonight's match is Manchester United v Wolves. No contest really as they represent the top and bottom of the Premier Division, so the 5-nil victory is hardly a surprise. I finish off today's paper which, being Sunday, is full of 'advice articles' - how to think more positively, how to play better golf and the importance of passion in everything you do. So I ask myself, am I passionate about what I am doing here? The answer comes back: I am simply happy to be here in this home with these good people, and that is all that matters.

I have a surprise the following morning when Anish (orphan number three) speaks to me for almost the first time. "Uncle, can I show you my project book?" Anish hopes to go into hotel management, so this is the subject of his project. He has written it very neatly, with illustrations, and is clearly proud of his achievement. I have a good look through and read every carefully written word.

"It is excellent, Anish," and I think that maybe I will be sponsoring him one day too…

✹✹✹✹✹

Fields on edge of city, Swayambunath Temple in background.

Roadside chess.

Reflection pool, temple in Patan.

Man selling Batik outside International Church.

Man preparing family for school.

Beating out copper bowls.

Philip Holmes' mosaic workshop.

Scenes from the Lalitpur festival.

(Left to right) Radheshyam, Arun, Nick, Anish, Indra Maya, Ashish and Sunil.

A visit to Chitwan National Park (we stayed at Sapana Lodge).

Sunil's college in Kathmandu.

Pushing taxi on family outing to Buddhist shrine.

The tailor Kiran in his new shop.

Ramesh with his brother Raj.

At the barbers (an amazing experience).

Nick with Mama Indra Maya.

Boys with motorbikes, resting on house-hunting expedition.

On the runway at Pokhara.

Our complete party of 13 on Pokhara trip.

View of Lake Fewa from our hotel.

(Left to right) Santosh, Niran, Ashish, Abhishek, Ramesh, Indu and Arun.

Cycling expedition round Lake Fewa, Pokhara.

Boating on Lake Fewa, Pokhara.

Fishtail mountain, Pokhara.

Sunrise from Mt Sarangkot.

Glowing in the light of the risen sun.

A rainbow appeared.

Memories of Pokhara.

# Chapter Eleven

Today I will be fulfilling my original purpose for coming to Nepal, which was to visit Manish at Triple Gem School. I had arranged today's meeting with Lama Kondan before I came away – so I hope it still suits him. The plan is for Ramesh to accompany me. This will be another new experience for Ramesh in Kathmandu, and he is looking forward to it. We arrive at the Swayambunath Temple easily enough, and a host of memories flood in about my teaching days in Nepal two and a half years previously. It takes some time to locate the school, but eventually we do, and I receive a warm welcome from the friendly doorkeeper.

"Mr Neek, how good to see you again! It is a long time since you were here. Are you coming back to teach?"

"Not this time!" I reply, laughing. "Let me introduce my godson, Ramesh Khadka."

We are shown in and climb up the familiar concrete steps to Lama Kondan's office on the top floor. And there he is, larger than life in his purple robes, sitting behind his desk.

"Nick! Nick!" he cries. "Come on in. It is so good to see you again."

Fortunately he is expecting us, and we are given the traditional silk scarves of welcome. I introduce Ramesh as my godson and they talk briefly in Nepalese.

"Now let me offer you a drink," he says, and we are glad of a Coke after our long hot journey. Lama Kondan is in terrific form, laughing and joking, as enthusiastic and energetic as ever.

"So, how is the school progressing?" I ask, because when I was here before he was having many problems with the parents.

"All that is over. Some parents took their children away and now we are going from strength to strength."

He tells us about the success of the school football team, their foreign trips, his first graduation class and the two hostels which have now become three. The school is clearly thriving and has a growing reputation. He shows us the new library and the new classroom for the top form, which has meant giving up half of his own study. There is a mathematical problem chalked up on the blackboard which intrigues Ramesh.

"This is a really interesting problem, the sort of equation which I enjoyed working out when I was at school. I would love to come and teach here in my free time."

We return to Lama Kondan's office and very soon Manish is shown in. He looks totally in awe of us, but settles down after we shake hands and sit down together to talk. He is only 12 and is reluctant to speak English except for 'yes' and 'no', but I am delighted to see him anyway. I ask him a few questions about how he is getting on at school and gradually he opens up a little. I say that I have been very pleased with his school reports, which I receive four times a year by email. He tells me that his favourite subject is Nepali, while his worst is IT.

"Now that we have met, Manish, I will be sending you an email occasionally. Would you like that?"

He nods and smiles. Break is over and he has to return to his lessons.

"I am very pleased to have met you," I say, "so let's keep in touch now."

"Yes, thank you, I would like that." He bows and runs off.

All this time, Ramesh has been talking to Lama Kondan, but it is soon time to take our leave. Later, Ramesh tells me how impressed he has been by this experience.

"Thank you for bringing me, Uncle. This is a school where I would definitely like to come and teach one day. Lama Kondan is a really good man."

We have been invited to lunch by Rajesh and his wife Soni, so we return to the main road where we catch a bus to Budhanilkantha on the northern edge of the city. The bus is so small and so full that standing passengers have to twist their bodies at odd angles because of the low headroom. I notice that Ramesh's right leg serves as someone's hand rail. It is a very squashed experience, lasting about twenty minutes. When we can disembark and straighten ourselves up, we see Rajesh waiting to meet us at the bus stop.

"Welcome! I am so glad to see you both!" he says. It is only a short walk to his house, which is surprisingly large. I learn that it has only recently been extended, but there is still a lot of work to be done. His beautiful wife Soni welcomes us in and we are offered fruit juice. The three of them have much to talk about and I think what a lovely family Ramesh has been adopted into. I have been told to expect a wonderful lunch because Soni has a great reputation as a cook. It is indeed quite delicious, with the addition of some side dishes which are new to me.

Rajesh wants to show me round. I gather that he has added two floors as well as a sun-terrace roof. Although the main structure is in place, there is all the plastering to do as well as electrical work and decorating. I only see one workman, so it is not exactly a hive of activity and I wonder why. The workman's wife is present on site, presumably to provide him with food and water through the day. But this will be a fine house when it is finished and from the roof there are spectacular views of the surrounding countryside.

"Come, I will take you for a nice walk. Soni and Ramesh have not seen each other for a long time and they will want to talk."

So again Rajesh and I find ourselves in each other's company, walking in the countryside. I can see why he has chosen to live here and I take in great lungfuls of fresh air, so welcome after the dustiness of Kathmandu. We climb up to a viewpoint overlooking meadows and fields of corn where the sound of birds combines with the bleating of goats. Being the sort of man he is, Rajesh occasionally waxes lyrical about the beauty of God's creation and he even bursts into song at one point. He is great company, full of chat, telling me how he wants to turn his home into an orphanage when it is complete.

"I would like to take in eight orphans, and we will be their Mom and Dad, and look after them so well. This is my dream, Nick."

He takes me on a different route home and shows me how extensively bamboo is used in this area: for fencing, for greenhouses, as well as for an entire restaurant. I also learn that bamboo is used to carry corpses to cremation sites.

The children, Sara and Basanta, have arrived home following their exams, and they show me their science and maths papers. I confess I would be hard pressed to achieve high marks in either exam. They show me examples of their classwork, which is beautifully written in near-perfect English. I should imagine they are model pupils. In fact, Ramesh tells me that Basanta reminds him of himself at this age: quiet, reserved and studious. If I ever return to Nepal, I would love to bring some English story books to read to them.

Soon it is time to go, but not until I have been issued with an invitation to come and stay.

"We will have such a good time together. We will have walks, and sing songs together, and Soni will cook you wonderful meals."

"I can't promise, Rajesh, but I would love to come again. Let's hope so. Thank you for today, which I have enjoyed so much."

There is one more call to make before we go home. This is to Indra Maya's family home in Pulchowk in the busy city centre, not far from Wasta Care Centre, where Karjun lives with his family as well as his brother Radheshyam. The property has a curious entrance. Ramesh and I are walking along a side street when we suddenly dive off to the right, down a dark low-ceilinged passageway which emerges into a complex network of stairways and curtained-off rooms. It reminds me of a warren, and exactly who has which bit I cannot be sure. I am led up two ladders which take me to a studio equipped with extensive sound-recording equipment and a keyboard. This is where all Radheshyam's creative work is done: composing and recording songs, mostly for church use, but also writing pop songs with Amit, the pastor's son.

Karjun is here as well as his two boys, Ashish and Abhishek, and we have a bit of a jam session. Word has got back via the family network that I particularly enjoyed one of the hymns that Rajesh sang to me last week in his church, so we all sing it together. Now I am invited to play something at the keyboard, because Ramesh has told everyone that I am an expert pianist. The machinery is set to record and everyone waits expectantly for the proof of my great talent.

This is slightly embarrassing as I don't have any music with me. Instead I try to improvise a harmony to another hymn tune they put in front of me and they all think it is brilliant. It isn't, but they are too polite to say so. It has been recorded and I am told that the recording will provide them with a cherished memory of my visit. Inwardly, I promise myself to be better prepared next time with some music of my own. I have loved coming here and meeting Karjun's boys again. Ashish is particularly musical, and says he would love to hear Adrienne play her violin one day, because that is his favourite instrument.

By now it is early evening and time to go back home. Ramesh and I say our goodbyes, but look forward to seeing them all the next day (which will be my last) when there will be a going-away party for me. I am very touched by this, but it highlights the fact that our time together is drawing to an end. It is night and we walk slowly home. It is a clear night and we stop for a few moments to look at the stars. I point out the constellation of Orion with its belt and silver dagger. I want to say to Ramesh how much I will miss him when I go home, but I can't bring myself to, so we carry on in silence.

After supper, I am given some photos of when the boys first came to Wasta Care Centre, and Ramesh helps me write their names on the back. Indra Maya presents me with a beautiful Nepalese shawl for Adrienne, and I tell her how pleased she will be. Ramesh has resolved to wake up early in the morning and take me for my last early morning walk. Knowing what a sleepyhead he can be, I have my doubts but we will see.

✱ ✱ ✱ ✱ ✱

# Chapter Twelve

Ramesh duly appears at 6:30 am, shaved and ready to walk – quite an achievement. He is evidently getting into student mode again, because while I am being taken to the airport later today, he will make his way to the bus station for the overnight ride to Lumbini. We walk up a steep path and eventually emerge into a smart residential area. I point out a poinsettia tree, some nasturtiums and two datura trees, the names of which he duly commits to memory. He meets a former teacher waiting at a bus stop and tells him about his medical training.

The day is warming up by the time we arrive home and, as a special treat, there is a box of Cornflakes on the table for my breakfast. The boys have never had cereal of any kind before and Ramesh looks into the box with deep suspicion.

"You must try some," I say to him, so he does – but only a tiny bit with just a drop of milk.

"Mmm, it's quite good," he says, but then returns to his favourite breakfast food: one piece of white bread, very thinly spread with pineapple jam, plus a small cup of weak, black coffee. He is not a big eater.

I am a little nervous about the family occasion which is to come, but I want to say a few words so I ask Ramesh to write out some sentences for me in Nepalese which I can practise. I dictate to him what I want to say which he puts into Nepalese. He then helps me with my pronunciation. I struggle with my soft and hard palatal 't's' and 'd's' and my 'chhu's' and 'chha's', but hopefully I will make myself understood when the time comes.

We have a second brief outing towards the end of the morning, but Ramesh then has to go to the bus station to buy his overnight bus ticket, so I return home by myself for lunch. A few members of the family have already arrived and by 1 o'clock we are a houseful. I think there must be about twenty people present, including baby Angel, Amosh's daughter. All the ladies are wearing colourful saris, and every seat in the living room is taken, with four people squashed on the bright yellow sofa. I know most of Indra Maya's sons, but not her daughter, Ira, or the various in-laws or grandchildren, so it is a little bit overwhelming. After we have eaten together, plates are cleared away and the hymn-singing begins, to the accompaniment of guitar, drum and tambourine – and this goes on with growing enthusiasm for about fifteen minutes. Then there is a time of silence before Mama begins to speak. Her son Rajesh, who has the best English, translates everything for me as she embarks on what she calls her testimony.

She relates how she was married when she was 17 and within a few years she had given birth to four sons and one daughter. Her husband then contracted tuberculosis and, at the age of 28, he died. (Rajesh told me later that, in her grief and agony she would pound her pregnant womb and so gave birth to her mentally and physically disabled son, Pancha.) As there was no pension for widows in Nepal in those days, she had to go out to work. This consisted of buying fruit and vegetables from the suppliers and growers early each

morning, and selling them for a few rupees at a tiny roadside stall. So the life of hardship began, and the children had to fend for themselves.

She explains how at one time she became very ill and could not work at her stall for several days. It happened that there was a school on the opposite side of the road which had been established by a Norwegian Church, and they made enquiries about her. They managed to make contact with her and on learning her story, they offered to employ her as a cleaner in the school. So for the next nine years, this is where she worked.

The Norwegian School wanted to start a small orphanage in Kathmandu, and knowing that Indra Maya had been a mother to six children, they asked her if she would be interested in being the house-mother to five orphans. A man called Leif Jansen would help her and be the main link with Norway. By now her own children were growing up and beginning to have families of their own, so she moved away from the family home and into an attractive house with a garden rented by the Norwegian Church. This was to be Wasta Care Centre, 'wasta' meaning 'nurture'. All of her sons would help her run the orphanage and look after the boys, and Rajesh would be in charge of their early education. Leif Jansen visited them each year to see that everything was running smoothly, acting as a sort of guardian to them; he would bring presents for the boys and take them on outings. But three years later he very suddenly died. The financial support continued but Indra Maya no longer had any personal support from Norway. Also she was having problems with the landlord who complained that she was using too much water for laundering the boys' clothes. Eventually they had to move – in fact they moved twice before settling into the ground floor rooms of the building where they are now.

As the boys grew older, she started to worry about their future, particularly that of Ramesh. He wanted so much to go to medical college, but how could she possibly pay the fees? She prayed for an answer, a miracle… and then one day two and a half years ago I arrived at Golgotha Church to join in the service, and she felt that maybe her prayers had been answered.

At this point Indra Maya looks at me; her testimony is over so there is a pause. Everyone has been totally quiet and respectful while she has been talking. Now they also turn to look in my direction and I feel I am expected to say something. Very slowly I feel my way into the subject, not quite sure where it will lead. Rajesh translates everything I say into Nepalese, as the family remain quietly listening.

"My first interest in Nepal came about when I saw a notice in a local magazine about a charity called "CHANCE for Nepal", founded by a lady called Barbara Datson. She was looking for sponsors for children attending Triple Gem School in Kathmandu. I responded to this and in due course became the sponsor of a boy called Manish…." And so I continue. After a few minutes, Ramesh arrives back and joins his family to listen to my story.

"…And now I want to help Niran continue his accountancy course in Delhi. I would like to thank you all for being so hospitable and friendly, and particularly thank Ramesh taking time off from his studies to be with me. I have had a wonderful ten days with you all, and I am sorry that they have now come to an end."

At Ramesh's prompting, I now say my six sentences of Nepalese – three for Indra Maya and three for everyone else. I think it is mostly okay, and of course they love it, giving me a big clap. I had written down ten things I had been grateful for on this holiday and I now ask Ramesh to read these out in Nepalese for everyone to understand. How accurately he does this I will never know, because he takes centre stage and has everyone in stitches of laughter, I suspect at my expense. He really shows his strength of character while he chatters on so confidently and fluently. What I wrote was not meant to be particularly funny, but he turns it into quite a comic performance and helps to dispel the rather solemn atmosphere.

That is more or less the end of the party. But I want to take a few photographs while everyone is together. So for the first time I find myself in the boys' bedroom, where various group shots are arranged for me. Ramesh then leaves to have his lunch, and I am on my own for the first time with Niran, Anish, Sunil and Arun. This is an ice-breaking moment for us as we have not really spoken together before. They are great fun and easy to talk to, and I find myself wishing we hadn't waited so long to chat together like this.

"I see that there is only one desk in your room."

"Yes, Uncle," says Sunil. "First it was for Ramesh, but now he is away it is for Niran."

"So this is the hot seat of study," I say.

"Yes – and when Niran flies off to Delhi, it will be Anish in the hot seat." They laugh out loud as they find the idea of a 'hot seat' quite amusing.

"Please let us email each other as I would like to learn more about you. So far I only know about Ramesh – and a little about Niran."

They gladly agree to do this, but then I realise they do not have computers. Maybe they will each need one in due course.

Soon it is time to depart. Karjun is going to take Rajesh, me and Mama to the airport, so I shake hands and give hugs to everyone. Finally, Ramesh and I have a big hug. This is quite emotional for us, and we are both a bit tearful.

"I am going to miss you, Uncle. This has been the best ten days of my life. I have never had such a good time before."

"I am going to miss you too, Ramesh. You have been a great companion and I feel proud of you. Thank you for everything you have done to make this such a great trip for me."

He cannot say *au revoir* as I had taught him, but manages *à bientôt* instead. There is a lot of hand waving as the car finally pulls away, heading for Kathmandu airport.

# Chapter Thirteen

A few days after my return, I received the following email from Ramesh:

*Dear godfather,*

*I am happy to know that you had a safe journey to your home. I am really lucky to have you in my life and I am again telling you that you provided me best ten days of my life.*

*I will be writing you more about my study and other interesting thing as I will have.*

*With lots of love and remembrance.*

*Your godson,*
*Ramesh Khadka*

Niran was soon to fly off to Delhi to take up his accountancy course, which left three boys behind in the family nest. So I now started to expand my correspondence to include Anish and Sunil (I had left them with enough money to buy a computer each). Gradually from their emails, I began to learn about the hardships of their early life and how they were taken into Wasta Care Centre, where at last they were given the love and care they so much needed. Their stories struck me as very similar to Ramesh's. Now they were in their late teens, they would also need help in choosing a career for themselves, becoming qualified and finding their place in the world. They began to confide in me and ask my advice, so I soon realised that I was going to be supporting all five boys. What confirmed this was a Skype conversation we had about six weeks after I had come home. Ramesh happened to be with them and one evening they gave me a call. It was an exciting moment to see each other again, and after I had spoken to them all in turn, Ramesh finally came on and said:

"So, uncle, you now have five godsons in Nepal!"

"Five godsons! Goodness me, how has this happened so suddenly?" but I realised it had and I felt very pleased. At that moment, I made an inward vow that, one way or another, I would give each of them the necessary support towards getting a good qualification. It was a big responsibility, but I knew this was what I wanted to do. They would be a great source of interest to me over the coming years, and I knew my support would mean a lot to them, too.

Towards the end of the year a new name appeared on my Facebook page: Santosh Karki. How this sort of thing happens I do not know, so I chose to ignore it. About a month later, his name popped up again, and then again early in the New Year. As it sounded Nepalese, I decided to reply by asking him who he was and how he had got hold of my name. This was his reply:

*Hi Nick.*

*Thanks for your message. I am from a simple family living in one of the village area called Palpa in western region. Recently at capital studying sciences at +2 level. My family is a poor background, so I must struggle hard to come up and doing so. Hope you reply me soon.*

He did not tell me how he got hold of my name, and to this day I don't know. But I was interested enough to reply so I told him a bit about myself, including my love of playing the piano. He replied saying that he would like to play the piano, but didn't know how. He also told me that he had been brought up a Hindu, but often found himself praying to Jesus and his prayers were always answered. He was interested in doing social work for the poor of Nepal, possibly in the field of medicine or education, but "I am financially weak". I also gathered he was 17 years old.

I wrote back saying that I might be able to give him some financial help and asked him what were his immediate aims. I also decided to ask Rajesh (the Pastor) to go and see him, which he promptly did. Rajesh must have told Santosh about Ramesh at the Medical College in Lumbini. Santosh's next email to me read:

*Thanks, millions of thanks. Till today nobody had given me this much of support in my life. Thanks to Jesus because he have listened to me. I have met Rajesh brother and he also have told me to get courage. I really feel lucky in meeting him. I also want to study medicine in Lumbini Medical College in the place where Ramesh has studied. I have shown my house to Rajesh brother today. Hope we will meet in the same house and talk.*
*Love*
*Santosh*

At some point in our exchanges, I told him how I first came to Kathmandu in October 2009 and taught for a few days at Triple Gem School. This email led to the most unexpected reply from him: he said that he had been a pupil at Triple Gem School and remembered when I came into his classroom as a special guest and had given them a lesson in English History.

To think that our paths had already crossed! I could hardly believe it. Of all the schools in Nepal, how incredible that we had both been in the same one at the same time! It was an extraordinary coincidence. I asked him about his home life and his reply came as something of a shock to me (he now started calling me 'Godfather'):

*Thanks for letter Godfather. I live with my uncle and aunt nearly 30 minute walks from Swayambhu temple on the top of hill.*

*I left my parents at a small age and meet with them once a year. I left my parents for good education in capital because we do not have good school in village. You know I had a critical condition in my childhood. I use to sleep in the floor, when aunt and uncle slept in bed. I was given no good clothes. My school uniform was dirty and unironed. Every friend used to tease me and once I remember even Kondan Lama sir had took off my shirt and was given to wash. You know Godfather my uncle and aunt hates me a lot till now, they even used to beat me a lot in my childhood. They have told me that I must leave their house after my next exam. Even we share same kitchen they don't chat with me anymore. I haven't told this even to my parents because I don't know how would they feel when I tell them this and what would be thing they could do for me (nothing except listening because they do not have any option except that). Many nights I have cried alone because there was no one to see and support. You are the first person in the world that I have told my story. Hope we will talk more when we meet.*
*Love*
*Santosh*

I said how sorry I was for him, and asked if he could think of a way out of the situation. He replied:

*Thanks Godfather,*

*Your inspiration and support has greatly helped me. You know Godfather when you give me this kind of support I really feel that someone is with me at the moment. After I finish my higher secondary exam, I am going to leave my uncle and aunt house. That is two months from now but I have no any idea for future because my family need financial support and I need education. Don't know what to do. You can share any idea if you have.*

*Love*

*Santosh*

A few days later, he told me that a cousin of his living in Kathmandu had offered him a room in his flat, and he seemed happy with this for the time being. I wrote back:

*Hi Santosh*

*I was really pleased to hear that you are moving out today. It is sad that your aunt and uncle are so unfriendly towards you, and also that your own family cannot help you at all. I cannot imagine this situation because I always had so much love from my family. So I hope your cousin will be friendly and supportive towards you. Meanwhile, try to think positively about this start of a new life, and put the past unhappy times behind you.*

His reply came a few days later:

*Godfather Nick*

*You are the only one in the world to inspire and support me in my critical condition. I just have come to my cousin's room and I am tired (mentally and physically). I think that what I did in this situation is right thing, because I had no other option than this. I have felt a lot of support from you Godfather. Hope our relation will*

*last longer and forever. I am praying to Jesus to support me in my life, hope that he listens to me.*
*Love*
*Santosh*

*Dear Santosh*

*I have just read this, so now rest and relax. I can understand how you must feel, like being rejected and thrown out, and this is depressing for you, and you are so young really and need support at your age. I think you must have some friends that you can share your suffering with, as well as your cousin, but it cannot be easy, and you have all my sympathy.*

*But you have a place in the world and a perfect right to happiness so try to stay focussed on what you want to do with your life. Your life is what you make of it, and I think you have a strong mind and much determination.*

*Try to keep cheerful, Santosh, and remain always hopeful. I know it is a really tough time just now, so don't be too hard on yourself. Above all, keep calm and take things steadily. I am with you in this, so try not to worry.*
*Lots of love,*
*Godfather Nick*

He replied:

*I do not have words to say thank you, but I really thank you from my heart and soul. You are one real friend, father, teacher who supported me in this stage. We cannot find a person really supporting and kind hearted in these days, but I think Jesus made me lucky to get you. Yes, today I am bit tired. so promise you to write tomorrow. Good bye.*
*Love*
*Santosh*

I had told him about my forthcoming visit in March, so by now we were looking forward to meeting each other. I sent him some money to help him out in the meantime and buy himself a laptop. In England we had been experiencing one of the longest, coldest and wettest winters on record, so the prospect of ten days in the warm sunshine of Nepal was quite appealing. I flew out on 28th March 2013, having arranged to be met by Rajesh and Sunil at Tribhuvan Airport, Kathmandu the following day.

# Chapter Fourteen

It was while waiting for my flight out of Heathrow that I began thinking about this double life I was leading, partly in the UK and partly in Nepal. I remembered how bereft I felt, flying out of Kathmandu a year ago, knowing that I would not see Ramesh and his family again for at least a year. Was I going to experience this sense of loss and bereavement again in ten days' time? Was it foolish of me to have struck up deep friendships with people who, at best, I would only ever see once a year? More importantly, could this time away possibly match up to expectations based on my previous visit? Would the novelty of entertaining an English guest have worn off and the family no longer be particularly interested in me? Why should these young men show any regard for me, coming as I do from a distant country and a different generation? Would I like Santosh, or would I be disappointed in him, or feel I could not trust him?

Putting such questions aside I resolved to have no particular expectations for this long-awaited trip. I would simply enjoy it as an entirely new experience and remain deeply interested in everything that happened, as if it were all a delightful dream from which I would wake in ten days' time, leaving it behind when I returned to my life at home.

My arrival is slightly inauspicious because there is nobody to meet me at Tribhuvan Airport. It is 6:00 pm, the sun has set and I am pushing my luggage trolley among the taxi drivers who naturally all assume I am in need of their services. I feel foolish standing there so conspicuously, and wonder what can possibly have gone wrong with the arrangements. After nearly half an hour I walk back into the terminal building and discover a room full of local people waiting for passengers from arriving planes. I go in and peer among the faces, hoping to recognise a family member. Eventually I spot Rajesh and Sunil deep in conversation.

"Rajesh! It's me!"

"Brother! You have arrived! We have been waiting for you so long."

"Didn't you see me come through Customs?"

"No, because the sign kept saying, 'QR 355 not yet landed'. So we were not looking for you among the arriving passengers."

So that little mystery is explained. The plane arrived perfectly on time an hour ago, but the airport signs have not yet caught up with the fact. Never mind – things can only improve. We laugh it off and are soon on our way to Wasta Care Centre. (Apparently, Mama phoned earlier to enquire why they had been waiting such a long time. She thought maybe there had been an air traffic jam in the skies overhead!)

When we arrive I am greeted by two of the orphans, Anish and Arun, Radeshyam the musician, the two sons of Karjun, Ashish and Abhishek, and of course, Indra Maya. How wonderful to see them all again, my family in Nepal! I find their presents – a green woollen shawl for Mama, wind-up torches for the adults and Easter eggs for the boys. The torches are not so essential now that they have solar power (this has been fitted since my last visit). During the daily power cuts (or "load shedding") they now have electric light in every room, which is such an improvement on the conditions here last year.

Of the five orphan boys, Anish is the most soft-hearted and solicitous towards me. I feel he is also the one most in need of encouragement, support and advice. His choice of doing a catering course is turning out well and now he is keen to show me something.

"Uncle, can I show you my presentation on cake?"

He produces a portfolio demonstrating the history of cakes, how to make a black forest gateau and other cake-related pieces of information.

"I had to give a talk to my class about cake, and my teacher was so strict, always interrupting me with questions, asking for more information. But it went well, I think."

He has prepared for me the traditional Nepali *dal bhat* supper: cooked rice, lentil sauce and a bowl of vegetables. It is delicious. "Great meal, Anish," I tell him, and he is delighted.

Later on, I have two short telephone conversations. The first is with Ramesh at Lumbini Medical College, and the second with Niran, studying accountancy in Delhi. They are sorry not to be with me for this trip, but they wish me a happy time at Wasta Care Centre and we will speak again. Soon after these calls, I say goodnight to everyone and collapse on my bed, ready to catch up on the night's sleep which I missed on the flight.

The next morning there is a gentle tap at my door at about 6:30. It is Arun.

"Uncle, would you like me to go for a walk with you?"

These walks before breakfast have become absolutely traditional and, just as last year, will be unfailingly observed. The only surprise will be which boy will knock on my door.

Arun is 17 years old and the youngest of the five. He is tall and willowy, very softly spoken, hesitant and retiring - a more harmless person I think one would never meet. Although he can speak and understand English, conversation is not easy because his voice is so quiet. I often find myself having to stop in my tracks, draw my ear close to his mouth and ask him to repeat himself. After a series of 'ers' and 'ums', we usually manage some form of conversation, but I am never fully sure that I have actually understood him. His loose sideways shake of the head (which in the West would indicate 'no') here means a definite 'yes'. The only problem is that 'no' is also suggested by a sideways but slightly different shake - and it is not easy to determine which is which!

Last year I arrived on 10th March, but it is now two and a half weeks later in the year and noticeably warmer so I will be wearing a T-shirt and sandals for the next eleven days.

I am by now quite familiar with the walks in their neighbourhood and if you ignore the litter and the smells, they are quite pleasant. Nakhu, the name of this area, is on the southern edge of the city, and is comparatively unpolluted. I suspect it was largely undeveloped a few years ago, but houses now seem to be springing up in a totally random fashion with little if any planning control. We walk across a small river (now provided with better stepping stones than last year) on to a flat grassy area, where groups of children play cricket and football. There are a few goats grazing. The walk takes us past a few small dwellings and the occasional shop. People are gradually waking up for the day, as are the chickens and dogs.

I gather from Arun that he has final exams quite soon and that he has decided to read science at college next year.

"Although I am not very good at science, Uncle," he confesses, "I would like to study engineering for my higher education."

I tell him I think this is an excellent idea and of course I will support him as much as I can.

Being Easter Saturday, the whole family goes to Golgotha Church, and as usual it is a full congregation (I guess at least a hundred) presided over by Pastor Shyam Nepali. It is interesting to see how many of Indra Maya's family are involved in the service: Radheshyam leads the music, Amosh preaches the sermon, Sunil leads the 'worship' section and Ashish plays drums, guitar and also sings. For a young man Sunil is particularly impressive. While everyone pours out their heart's devotion, he cuts an imposing figure of stillness and authority. He clearly has no qualms about being the centre of attention.

After the service I am introduced to some members of the congregation, including a local man who is a film director and only a very recent convert to Christianity. I am told later that he is extremely wealthy and is closely connected to the royal family. On Friday, he is going to be baptised at Golgotha Church.

I walk back home with Sunil, and he tells me how nervous he was about leading 'worship' because this was his first time, but he had practised it on Friday evening and thinks it went quite well. I tell him he appeared totally relaxed and confident - I would never have guessed it was his first time.

Rajesh has arranged for me to meet Santosh today. He arrives at Wasta Care Centre after lunch together with his cousin Krishna. This is quite an important moment for me, because of all the email exchanges I have had with Santosh. Both he and his cousin speak excellent English and they have cheerful, bright personalities. Santosh looks older than 17, and gives the impression of being an intelligent young man, confident and self-assured. Although there is a thirteen year age difference, they are apparently getting on well together in Krishna's flat. They take it in turns to prepare meals, and they enjoy each other's company. I ask Krishna what his work is.

"I work on *The Himalayan Times*."

"As a correspondent?"

"No, I am in charge of distribution. My section is involved in getting as wide a readership as possible."

I tell him how I used to read his paper every day on my trip last year. Ramesh would go and buy it for me and I thought it was very good.

Krishna is also writing a novel which he hopes will be translated into English by a professor of English he knows at university.

They do not stay very long, but before they go I tell Santosh that I will be seeing Lama Kondan later in the week and will then be in touch with him again. Santosh seems pleased to hear this and says he will look forward to hearing from me again soon.

After they have left, there is a long family discussion about the merits or otherwise of my sponsoring Santosh at Medical College. Mama is strongly against it. In her opinion it is too expensive and the money would be much better spent on helping other Nepalese orphans. Rajesh seems to agree with this view, but I'm not sure that Sunil does. I think he has taken to Santosh, partly because they are the same age and therefore at the same stage of their academic careers. I am still undecided, but I tend to think that Santosh is a bright young man who deserves the same opportunity that I am giving to Ramesh.

Anyway, the matter is closed for the time being because we want to make an expedition to the annual Lalitpur festival, which is only a short drive away. Eight of us set off by taxi and motorbike: Mama, Anish, Sunil, Arun, Rajesh, Ashish, Radheshyam and me. The festival has completely taken over a few streets which are thronged with local people. Once parked, we set off on foot, while trying to keep track of each other in the crowds. Anish holds my hand most of the time for fear of losing me. I take some pictures of Newari ladies dressed in traditional costumes, displaying craftwork and making food. Two women are demonstrating how a complete tangle of straw can be turned into a neat pair of shoes; another two are making wicks for oil-burning lamps; others are filling thin strips of paper with spices, then twisting the paper into tapers for burning as incense in Hindu temples. We wander on through the streets, jostling among hoards of people wherever we turn.

Eventually, we arrive at Hanuman Dhoka Palace Square, which is the home of the Kumari Devi, the Living Goddess. The custom of worshipping a prepubescent girl as the source of supreme power is an old Hindu-Buddhist tradition that still continues to this day in Nepal. We join the crowd gathered in the small courtyard. There, behind a grille, sits a young girl of about 12, bestowing blessings on the faithful and daubing their foreheads with red paste. She wears a mournful expression while she mechanically carries out her task, her beautifully manicured hands totally besmirched with the paste. The Kumari's 'godhead' comes to an end with her first menstruation, because it is believed that on reaching puberty the Kumari turns human. But meanwhile the poor girl has to continue performing her daily rituals for the benefit of her worshippers.

We continue our tour for an hour or so, and see dancers, musicians and craftsmen all contributing to the festive atmosphere. Anish continues to steer me until at last we are back where we started. The decision is now made to go back to the family house at Pulchowk. Here we will do some singing and playing together as we did on my last visit. It is growing dark as we weave our way round the streets, arriving eventually at the little side road where we all disappear down the narrow alleyway which leads into the old family house. As before I am led through tiny rooms and up two ladders until we arrive at Radheshyam's recording studio where we squash in together sitting either on chairs or on the floor.

The main interest for me is to hear Ashish, Karjun's eldest son, singing his own recent composition. He announces it as a devotional song about the Day of Judgement. He has a sweet, mellifluous voice and sings with considerable feeling and expression.

I have offered to take Mama, Anish, Sunil and Arun out to a restaurant. It was to have been Kentucky Fried Chicken but they decide it is too far away, so instead we go to the Road House Café, which turns out to be a good choice. Anish, the one pursuing a career in catering, takes command of the situation. He recommends dishes from the menu, places the orders and makes intelligent comments on the quality and service of the establishment. I have spaghetti carbonara while the others have pizzas - to follow we all have tiramisu. This is a very enjoyable occasion, and I feel I am following in the footsteps of Leif Jansen, their former 'godfather', who used to take them out to restaurants occasionally on his visits to Kathmandu. There is an embarrassing moment at the end, however, when I realise I don't have enough money to pay the bill. Sunil has a quiet word with Mama, who discreetly disappears to the toilet with her handbag and comes back with the necessary rupees. Nobody seems to mind too much and we leave the restaurant at about 9:00 pm. And there is Karjun with his taxi, waiting to drive us home - as is the way with this family, the mobile phone network operates so that we don't have to wait a minute for a lift – it just happens!

Back at the orphanage, Sunil shows me some of the letters of instruction he sent out to students during the year he was President (or Head Boy) of his school which is called, rather surprisingly, Oxbridge College. This has been an important year for him and he has found himself enjoying the power and authority that go with the role. He is a fearless, independent boy who likes dabbling in local political issues. He is confident standing up in front of a crowd, whether to make a speech or to lead worship at church. He is ambitious and has a plan to come to England one day to read Business

Studies. In some ways, even though he is still only seventeen, I think he is the most self-assured of the five boys.

# Chapter Fifteen

The next day is Easter Sunday and at 7:00 am there is a knock at my door.

"Come in." It is Sunil.

"Morning walk, Uncle?" and he does a few running steps to suggest it is going to be an energetic walk. The mornings are still slightly cool but I don't bother with a jacket knowing that in an hour the sun will be up and the temperature climbing quickly.

We set off briskly and on the way he tells me how he became President of his college:

"When the time of the election came round, I never thought about putting my name forward, but one of my teachers told me I should, so I did. And in fact I won the election easily. To start with I was nervous about all the responsibilities, but I soon found myself enjoying them: organising the programmes, making speeches, issuing letters to the students about dress code, tidiness, haircuts etc. And of course I have made a huge number of friends over the year - students who respect me and come to me for advice over disputes and arguments. So I have often to act as a judge in these situations, which can be difficult, but I enjoy the respect they show me."

Looking ahead, Sunil sees himself as the director or manager of some business enterprise. He seems quietly confident and we talk more about the possibility of his coming to study in England. He seems convinced that the best courses are in England and will do some investigation after I have left. Later in the week I will talk to the head teacher at Oxbridge College.

But for now, we have to get ready for a three day trip. Rajesh, Sunil and I are going to catch a plane to Chitwan National Park in the south of Nepal, the region known as the Terai. From there we will travel by bus to Palpa in the west to visit Ramesh for a short time at Lumbini Medical College. Finally, we will visit the World Heritage Site at Lumbini, the birthplace of the Buddha, which I have already visited with Ramesh and which now Sunil wants to see. But part of the excitement for him is going to be his first aeroplane flight.

We pack a few things and when Rajesh arrives later that morning we take a taxi to the airport. My only qualms about this trip concern the notorious safety record of Nepal's internal aeroplane flights. So Rajesh has booked us on Buddha Airways, which he assures me is very safe.

The forty minute flight to Bhairawa turns out to be not only safe but enjoyable, and Sunil takes endless photos on his phone both out of the plane and in the plane itself. We have great views as we fly over the southern rim of mountains forming the Kathmandu Valley and then the flat plains of the Terai.

We are going to stay at the Sapana Village Lodge Hotel, built beside the Rapti river in the homeland of the Thoru, the name given to the people of the jungle. The hotel itself was built by a man called Dhurba about nine years before, on a piece of land he acquired on the edge of Chitwan National

Park. His story is an interesting one. He was one of ten children born to a very poor family in this area and as a boy he attended Shree School, which is five minutes drive from Sapana Village Lodge. By the time he was 15, he had left school and was working as a waiter in a local restaurant. One evening a Dutch family came to dine there and there was something about Dhurba's bright energetic manner which attracted their attention. They asked him over to their table.

"You look an intelligent young man," they said. "What do you need most in the world to fulfil your dream?"

Dhurba thought for a moment, then replied, "If I had $5000 I could buy a piece of land and build a hotel on it. I would offer elephant rides and safaris to Chitwan National Park and ensure the protection and wellbeing of this lovely environment."

The Dutch family listened with interest but made no promises. But a few months later, Dhurba received a cheque for $5000 through the post. There was a note attached: "Here is the money you said was needed to fulfil your dream. You can either spend it on a few wild parties, or you can live out your dream. The choice is yours."

Dhurba of course set to work building his hotel, giving it the name 'Sapana' which means 'dream'. Over the course of nine years he has created safari-style lodges out of local materials alongside a central restaurant and office complex, where visitors can meander through gardens, wash down the resident elephant, relax in hammocks and enjoy some gracious living in a peaceful setting on the edge of the National Park.

We arrive by taxi at midday and are shown to a room on the first floor of our spacious two-storey lodge. I am surprised to see that most English flower, the dahlia, growing in wild profusion in all the beds. After settling in, we go for a light lunch in the restaurant and await the arrival of Dhurba.

Our visit here has been prompted by my English friend, Barbara Datson, founder of the charity 'CHANCE for Nepal'. She had been invited to stay here the previous October, and, while talking to Dhurba, became interested in the local school, Shree Secondary School, which she learnt Dhurba had attended as a boy. One morning Dhurba took her there and introduced her to the Principal and other members of staff. Being always charitably disposed to the Nepalese, Barbara asked if there was any way in which she could help this poorly supported government school (apart from paying the teachers' salaries, no state funds were available for anything else such as stationery, meals or playground equipment). When she made this offer, one childhood memory immediately struck Dhurba:

"I remembered the awful gnawing feeling in the pit of my stomach when midday came round and I had nothing to eat. How wonderful it would be if Barbara could finance a midday snack."

The Principal agreed with this idea. One of the problems at the school was that the children often went home for lunch and then did not return for afternoon lessons. If the school could provide them with a midday snack ("tiffin" as he called it), then this problem would be overcome. So Barbara agreed to finance tiffin for the children of Shree Secondary School and wanted me, six months later, to check that the system was going well and, if possible, to obtain some receipts.

After lunch, we meet Dhurba, and I congratulate him on his lovely hotel.

"How many staff do you have?"

"Twenty-two."

"And are you pleased with how it is going?"

"Very pleased. Visitors come here from all over the world."

As he drives us in his jeep to the school, he points out two pieces of land on either side of the road.

"I have bought both these sites. On this one I will be building a school and on the other I will build a training institute, dedicated to the environmental needs and sustainability of the National Park."

There is no stopping him!

At the school, we are unfortunately too late to witness the partaking of tiffin but are assured the scheme is a total success and that afternoon school attendance is now much better. I am given the receipts Barbara requires, a handwritten letter of gratitude from the Principal and even shown the tiny shop near the school gates from where the owner manages to provide a small meal for the modest cost of 16 rupees (about 12 pence) per child.

With our mission accomplished, we consider ourselves free to amuse ourselves for the rest of the afternoon. The obvious thing to do is to take a ride on an elephant through the jungle. We make a booking and an hour later find ourselves climbing into a howdah on a 40 year old elephant and settling down for a two-hour ride through the National Park. It's a curious lolloping sensation riding on an elephant – it's like taking one step forward and half a step back – but after a while you get into the rhythm of it and begin to enjoy the gentle rocking movement. Our eyes are trained for wildlife in the forest and we see several deer, a pair of exotic woodpeckers, a kingfisher and a white rhino, so we feel we have done well.

We are driven by jeep back to Sapana Lodge where we have tea and later on supper on the restaurant veranda in the light of the setting sun. To end our leisurely day we lie in hammocks under the stars. We begin to discuss Sunil's future and his wish to study in England one day. Well, why not? I think – we can talk about it more at his college when we get back.

The next day we move on to Tansen in order to see Ramesh at his Medical College. It is a long bus ride to the town, where Karjun, Ramesh and I stayed last year, but this time we decide to stay at a different hotel – an unfortunate decision as it is neither very clean nor hygienic. We go back to the same restaurant, the Nanglo West, for lunch and here we have an interesting encounter. While we are enjoying our *dal baht* and some rather chewy mutton, a young European man sits down at a table near ours, opens a local newspaper and proceeds to read it. Soon the waiter arrives to take his order which he delivers in perfect Nepalese. I am so intrigued by this that when I have finished my meal I go over and introduce myself.

"I couldn't help hearing you speaking so well in the native language. I am Nick from England and these are my friends from Kathmandu. Please tell me, what are you doing here?"

Naturally he also speaks perfect English.

"My name is Gunnar Mollestrad and I come from Norway. I have been working at the Tansen Mission Hospital for three and a half years, which is why I am able to speak a little Nepali."

He tells me more about his background and then, when the waiter brings his *momos*, I proceed to tell my story. He is interested to learn about Ramesh and he gives me his email address to pass on in case Ramesh should ever need advice about his future.

"Tell him to look into Post Graduate Medical Exchanges. Canada and America do them, so maybe the UK does as well."

We spend the rest of the afternoon booking our flight home from Bhairawa until it is time to catch a bus back down the hill to Lumbini Medical College. I am delighted at the prospect of seeing Ramesh again, even if only for a couple of hours. When we arrive, he is waiting for us at the entrance to the college, looking thin but smiling, with his hair buffed up in a new stylish way. He is wearing his smart college T-shirt and altogether cuts a neat figure.

After many hugs and greetings, he says, "I will take you on a short tour and then we will go to my room."

I am surprised at the amount of building going on here, even since last year. Clearly, the Principal is doing an effective job. This time I notice a higher standard of finish to the new departments: attractive flooring, good quality furniture and expensive-looking equipment. I am struck by the notice I read that Lumbini College is aiming to achieve a high reputation in Nepal as a "Medical Centre of Excellence". Ramesh is obviously proud of his college and makes a very good guide.

When we approach the male student hostel, he stops. "Do you see those bars on the outside of all the windows? They have recently been installed because boys were climbing out of their windows, walking along the ledges and into the next room."

"But why would they do that, when they could just as simply walk down the corridor?"

"I don't know, Uncle," but he is highly amused.

He takes us into his small room and we sit down where we can. It does not, of course, really cater for guests. Also half of it belongs to his room-mate, Sabindra Tandukar, with whom he has shared from the beginning and will continue to share until the end of the course. Thick medical volumes are piled on his desk with 'RAMESH KHADKA' printed indelibly across the outside pages of each one. The room leads onto a small balcony which affords a good view of the grounds, sports fields and distant hillside.

"Do you sit out there much?"

"Well, it is nice when it is dry. But when it rains there is nowhere for the water to go and if you look up you will see a gutter which pours all its water straight down. I had thought about complaining, but then I could not really think of any solution. So we just put up with it - and get wet!"

He tells us his daily schedule: study and lectures begin at 8:00 am and continue until 4:00 pm, with a break for lunch. Then he has some time for recreation, perhaps a game of football. This is followed by a snack and perhaps a nap until dinner at 8:00 pm. After that, his own private study time begins in earnest, lasting until 1:00 or 2:00 am.

"Really, there is so much to learn and I cannot possibly take it all in. But I use highlighters on the important pages," (he shows me some examples), "and that is the best I can do."

"How did you get on in last year's exams? You never sent me the results."

"I am sorry, I will do that. But I did well, better than in the first year, so I was really quite pleased."

"Is there a big drop-out rate here?"

"Not exactly, but many students have to go back a year if they fail their exams, which means the five-year course can become quite extended for them."

"Are you ever worried that might happen to you?"

"Absolutely not! I mean to work hard enough so I graduate in the quickest possible time. I do not want to waste time like that!" I could see he meant it.

I mention our meeting with the Norwegian doctor and he puts the name and contact details in his phone. Then I ask him if he has been keeping well.

"You know, Uncle, I have been having some problems with an echo effect in my right ear. I have just started a course of antibiotics for this, so hopefully it will improve.

Also I seem to be having dizzy spells sometimes, but I don't think they are too serious. The worst time was after my last overnight bus ride from home. I was sitting in the back of the bus, where it is so bumpy and I could not sleep all night. When I arrived at college the next day, I felt so terrible and it took me some time to recover."

"Why don't you take the aeroplane to Bhairawa, and then it is only a four hour bus ride as opposed to ten hours?"

"But that would cost about four times as much!"

"Well, at least you could take the flight one way. It does not matter so much when you are going home, but when you come here for a new term you must be fit and well."

He looked distinctly undecided about this, so I say, "Look, I am going to insist. You must come by plane, understand? You cannot refuse me if I insist, can you? I will want to know that this is happening – all right?"

"Okay, Uncle," he says a bit sheepishly and, smiling, shakes his head to indicate 'yes'.

He offers to buy us tea at a small hotel near the college entrance, so we retrace our steps, meeting and greeting several of his fellow students on the way. He chats to Sunil and Rajesh for much of the time, while I am happy simply taking in all the impressive surroundings. Lumbini Medical College is fast developing into a huge educational campus, but there is plenty of room to expand without in any way spoiling the beautiful rural landscape in which it is set. Again I am pleased that Ramesh has found such a conducive environment for his medical training.

The tea is very welcome, but the room is slightly noisy, so while Sunil and Rajesh are happy sending and checking messages on their phones, I monopolise Ramesh for our remaining hour.

I have brought a number of things for him which I had promised over the past few months: a recording of Handel's

*Messiah,* which had belonged to my mother; the DVD called *Orbit* about the Earth's annual revolution of the Sun which I had promised him earlier; the programme and poems from my recent mosaic exhibition; and one or two bits of writing which I thought might interest him. We talk a little about his future and I tell him that Sunil is interested in studying in England and that I have been to see his college. He enquires after my health, about my trip so far and how we are going to spend the rest of our time together in Kathmandu. Rajesh and Sunil now join in the conversation and we enjoy our last few minutes together.

Twilight is gathering when we step outside to order a taxi. We say our farewells and Ramesh gives me another big hug. The thought rushes through my mind that I may never see him again, life being so uncertain, but it is no time for sad thoughts and we are simply grateful to have spent this time in each other's company. We promise to speak on the phone before I leave the country.

Back at our hotel, we delay going to bed for as long as possible, fearing attacks by mosquitoes and suchlike, so we chatter far into the night. But like all things which you dread, the hours of night pass and sleep does eventually embrace us, leaving us fresh for our journey to Lumbini the next day.

We have booked a taxi to pick us up at 7:00 am and we are glad to quit our hotel as early as possible, avoiding any offers of breakfast. Instead, after a couple of hours, we stop at an appealing roadside cafe where we enjoy an appetising meal of deep fried pancakes, spicy vegetables and milk tea. We arrive at the World Heritage Site just before midday and the temperature is soaring. No chance of a downpour this year! We hire bikes and set off on our tour, the first stop being the famous birthplace building itself. I am happy to sit quietly under the trees and prayer flags while Rajesh and Sunil tour the building. This for me is the best part of the day, to be sitting quietly in this sacred spot, enjoying its peaceful atmosphere. There is no mistaking the special quality of this place; I felt it last year and I enjoy again now.

When they return, we walk slowly back through the gardens, past the pool and the pillar of King Ashoka and go to collect our bikes. For the next hour or so we cycle in very leisurely fashion past the great lake and through the complex of modern Buddhist temples which Ramesh, Karjun and I had visited in pouring rain last year. Quite honestly, those conditions were preferable to this overpowering heat and soon I am asking if we can leave the site and return our bicycles.

In an hour we find ourselves at the Buddha Bhoomi Hotel, where we stayed last year. Just as before, they provide us with an excellent lunch and, for a small additional fee, a room for the afternoon where we can shower and rest before catching our plane from Bhairawa back to Kathmandu. What a welcome time this is! By 4:00 pm we are rested, washed and refreshed for our flight home at 5:30.

Buddha Airlines once again do us proud, this time even handing out drinks and newspapers. Sunil must have taken over a hundred photos (including several of the air hostesses), and I think he, like Ramesh last year, has enjoyed this trip away. He has been entertaining company with his teasing comments and infectious personality, and it has been a good opportunity for me to get to know him better.

We receive a warm welcome home at Wasta Care Centre from Anish, Arun and Mama. Anish has prepared a lovely dinner and while we enjoy this, I tell him about Sapana Lodge Hotel. Anish has a dream to establish his own restaurant one day, so he is excited to hear about this place.

"I would like to visit Sapana Lodge some time with Rajesh and Ramesh. It sounds like the sort of place that will give me ideas for my own restaurant."

I tell Mama (through Arun as interpreter) about our meeting with Ramesh. I say that he is looking rather thin and is complaining of ear trouble and slight dizziness, but overall I think he is quite well and enjoying his time at Medical College.

Now, I am a little concerned about Sunil's English accent. I would like to try to improve it because it is often difficult for me to understand what he is saying. He reads out a short story for me and basically the problem is that he speaks too quickly. I subsequently learn that his own family and friends have the same problem understanding him because of his rapid delivery. Perhaps it is a habit he will outgrow in time.

✸ ✸ ✸ ✸ ✸

# Chapter Sixteen

Today Sunil takes me to visit Oxbridge College where he has held his presidency for the last year. It is by far the most impressive school I have seen in Nepal with well-equipped classrooms, spacious sports grounds, a high standard of decoration, good facilities such as social areas and a canteen, as well as some attractive plants and shrubs in pots. I meet the headmaster who is a cheerful genial man with excellent English. He himself came to do postgraduate studies in England and said he had no problem obtaining the necessary visa. He will obviously be able to point Sunil in the right direction if he decides to do the same thing one day. He speaks about how impressed he was by the high level of courtesy and good manners he experienced in England. We talk about the 18th Century Enlightenment and the Age of Reason, and share some of our enthusiasm for English Literature. I congratulate him on his school and he says he has been pleased with Sunil's year of office. This has been a worthwhile visit for me and I feel particularly proud of what Sunil has managed to achieve here.

A few days earlier, I had mentioned to Sunil that I might be prepared to buy motorbikes for him and Anish. This idea has now taken firm root and developed further, with the result that I am now going to buy three motorbikes, one for each of them and one for their cousin Ashish (to replace his scooter). So this afternoon is given over to a tour of some second-hand motorbike shops. The level of excitement in these teenagers' hearts and minds is palpable as they climb onto bikes, put on helmets and imagine themselves roaring through the streets of Kathmandu. It is quite clear that they are living out long-held fantasies and now at last the reality is within reach.

They are going to have to be patient, however, because at no motorbike shop can I use a VISA card – it's "cash only" – and I cannot access the large amount of money needed from the ATMs. So they will have to wait until I can send the money from England. As consolation, I promise to buy them brand new Pulsars (the bike of their choice) rather than second-hand bikes. They are thrilled about this and more than happy to wait for the money to arrive.

After lunch, I set off with Sunil again. This time we catch a local bus to 'Elephant Park', an attractive area on the outskirts of the city which he wants to show me. The day is particularly windy and the dust from building works on all sides flies into my mouth and nose, giving me breathing problems. We stop for a drink and then set off again, back through the crowded streets. I want to set Sunil up to do a computer course in his summer months after college, then buy him a new pair of shoes because his are uncomfortable and almost worn out. We manage to do both these things and head for home late in the afternoon. The wind has picked up considerably creating dust storms which obscure one's vision and make walking both dangerous and uncomfortable. At the crossroads near their home, the traffic policeman is having a fairly miserable time of it. He is standing in the middle of the road, not only battling against traffic chaos but also the dust clouds which are obscuring his view. Occasionally he all but disappears in a complete cloud of dust. Sunil guides me gingerly across the road and down the steep hill opposite, where more dust is being created by building works which are taking place to rebuild the wall of a local cemetery. A team of workmen are busy smashing rocks with sledgehammers while others assemble them into a rough wall. There is nothing the local labour force likes to do more than create a lot of dust, and on a windy day like this they go at it with extra gusto.

So I am relieved when we reach the comforts of home and am able to wash and brush myself down, relaxing with a cup of tea. The rest of the afternoon is spent watching a hilarious Korean film about a female cop and her hapless male victim. After a supper of noodles and cucumber, we have a time of worship, hymns and prayers. Then I play and sing through some hymns I brought from home, as well as the songs from Andrew Lloyd Webber's *Joseph and his Amazing Technicolor Dreamcoat*. Finally, Sunil gives me a wonderful head massage, while I listen to the others who read me stories in English for my entertainment. If I ever worry that I am spoiling this family, I need only recall the times when they are spoiling me. Unfortunately, Sunil insists on ending his head massage treatment with a complete rearrangement of my hair which makes me look like a cockatoo. They all find the results hilarious – but I utterly forbid the posting of any photographs on Facebook.

My walking companion the following morning is Arun's older brother, Anish, who has low self-esteem and is slightly prone to moods of sadness, even despair. This is perhaps hardly surprising when one considers the abuse he endured as a child: beatings from his father, lack of food, lack of care or parental affection. He wrote to me once saying that when he was a child he used to ask God why He created him as such a useless, unwanted person.

Today he takes me to visit the first two locations of Wasta Care Centre. The original orphanage was a two-storey house set in an attractive garden with a lawn, flowers and bushes and bordered by mature trees. Here the five boys spent their early years together, being taught by Rajesh, cooked for by Karjun and Radheshyam and playing basketball in the garden. These were happy times, marred only by the constant nagging of the landlord, who was always complaining that they were using too much water. His continual harassment in the end drove them to find other accommodation.

With the help of their Norwegian guardian Leif Jansen, they managed to find another house less than a mile away, which was roomier but had a much smaller garden. To start with it was a good move and at least they were not harassed. The problems began about two years later when the water supply began to turn foul, as it did for other houses in the neighbourhood. It became so bad that the family was forced to take their laundry down to the nearby river, which in those days was reasonably clean and unpolluted (now it is filthy and all but dried up). Representations were constantly made to improve the situation but nothing was ever done, so in the end the family was forced to move to the much smaller accommodation where they are now. By this time Leif Jansen had died, so life was becoming difficult again. They were still supported by the Norwegian Church but missed his personal involvement and friendship.

While Anish is showing me this second house, it still looks empty even after all these years, so we suspect that the water problem has not been resolved. It has been an enlightening tour and Anish has been the kindest and gentlest of walking companions. But before we go home, he wants to take me to meet a friend of his, a Finnish lady called Seija whom he met through a fellow student at college. This is turning out to be rather a long pre-breakfast walk, and in fact we have quite a lot of climbing to do before we arrive at her flat. But it is worth the effort. An elegant lady of about 70, Seija gives us a warm welcome on our arrival.

"Come on in. How nice to see you, Nick! Anish has told me so much about you. We were expecting you for tea yesterday, but I understand you got held up."

I apologise for this, explaining that the dust storm exhausted me and I really did not want to go out again after my shower.

"Never mind, you are here now. My student lodger Binod will make you some tea and you must have some of his chocolate cake which he baked for you."

I am introduced to Binod, Anish's friend from catering college, and while he and Anish prepare the tea, she tells me about herself. She and her husband have lived and worked as scientists in Finland all their lives, but in recent years have been involved in water projects in rural Nepal. As a result, they now divide their time between the two countries. Soon, she will be going back home because the summer months are so oppressive in Kathmandu. She also runs Alpha courses and is a regular attender at the Kathmandu International Christian Church where I took Ramesh last year. So, with my own interest in the charity WaterAid, we find we have a lot to talk about. She speaks very good English and I enjoy hearing her stories about living in Nepal and, in particular, Kathmandu.

"I like keeping open house," she tells me. "Across the compound there live some Hindu families. When the parents have arguments, their children run to me for comfort and a smiling face. People drop in and see me at all times of day and night, so you are also welcome here any time."

I think Anish also finds her home a place of refuge and consolation. She is a sweet motherly figure and I am glad he finds a warm welcome here. We return for a very late breakfast, but of course it doesn't matter. I have found that the passage of time has little meaning in Kathmandu. Because of the many problems one encounters in this city, it often amazes me that anything happens at all. Traffic hold-ups in particular account for hours of wasted time, so there is little point in trying to make or keep to appointments. Days simply turn out in their own way while planned events take place in a somewhat haphazard fashion. So a late breakfast is really quite immaterial.

Today Rajesh is going to take me to see Lama Kondan at Triple Gem School. Apart from wanting them to meet each other, my main reason is to find out what Lama Kondan remembers about Santosh Karki and whether he thinks I should be supporting him. By motorbike this is an easy journey, and once we are off the main road, I enjoy riding round the base of the vast Swayambhunath Temple which leads to my home territory of nearly four years ago. There is no school today but we find Lama Kondan seated behind a desk in his new office located on the ground floor. He seems bigger and more avuncular than ever.

"Nick, come on in. How are you?" he booms and gets up to give me a big bear hug. I introduce him to Rajesh who is given the same warm welcome. We have a long chat about his school and its progress, Rajesh's orphanage project, the five boys at Wasta Care Centre and what I am doing on my current trip. This brings me to the subject of Santosh. Lama Kondan nods gravely, knowing my interest because I have already emailed him about the boy.

"Let's have some tea first," he says, and he makes a phone call to order some. Then he starts, "Yes, I remember Santosh

Karki. He came here in Year 5 when he was living with his aunt and uncle nearby. His uncle was an army man, who believed in a strict disciplined upbringing and may have given Santosh a hard time, but I really can't vouch for that. He was a bright boy who worked hard and I remember that he shone in all subjects. But the first thing I must tell you is that he was then and is now still receiving support."

This sudden announcement comes as a terrible shock to me. I can hardly believe it. How can this boy, after all our email exchanges, not have told me? How can he have been so deceitful as to keep this information from me? All at once I feel totally deflated and undermined. Lama Kondan goes on:

"We get several visitors here, as you know, and one day a few years ago a man from Hong Kong arrived to visit the school. He took the time to talk to several of the pupils in the top class and, for some reason, took a particular interest in Santosh. He asked Santosh about his family circumstances and when he learnt about his troubled background, decided to take him under his wing. He paid for his education here and as far as I know is still supporting him. Even if that assistance has now come to an end, it seems that Santosh has not been entirely straight with you."

I find it slightly difficult to speak. I try to look Lama Kondan in the eye, but it is with a slightly trembling voice that I say, "Santosh told me nothing about this," but refrain from adding that it seems I have been taken for a ride, which is how I feel. Fortunately, Lama Kondan continues talking.

"So my advice, Nick, is that you need to be very careful how you spend your money. You, like Barbara and like many other Westerners I know, are very generous good-hearted people, but you need to think about the causes you are supporting. Why should Santosh be receiving maybe the best part of £30,000 or £40,000 from you to be trained as a doctor when there are so many orphans and orphanages like Rajesh's where financial support and help is badly needed? Santosh is a bright boy and a good boy at heart, but he is very young and will take anything that is going. He can look after himself. Speak to him again and establish clearly exactly what is going on. But there are many better causes to support than his, I suggest."

I feel so deflated that I can barely respond. Rajesh and Lama Kondan have plenty to talk about in their native tongue, so I leave them to it. But it is soon time to take our leave and I thank Lama Kondan for his helpful advice. He grins broadly and says, 'Don't take life too seriously, Nick', which is probably good advice. It had been our plan to visit Santosh's uncle and aunt to corroborate his story, but before we get on the bike Rajesh asks, "So we don't need to visit them now?"

"No, I don't think so."

On the journey home I feel both foolish and somewhat crestfallen. I decide that I do not want to raise the subject of Santosh again and probably Rajesh will let the matter rest now as well. Will I try to arrange to see him again before I depart? I don't know - but for now, I simply want to put the whole incident behind me and treat it as an unfortunate experience. I have acted against everyone's advice and feel foolish as a result. Oh well, not too much is lost and at least I feel I have done Santosh some good along the way.

# Chapter Seventeen

In my recent talks with Arun, I have learnt that he has never spent a single day outside the Kathmandu Valley. In fact, he has never had a day out anywhere. Well, all this is soon about to change. Rajesh has invited us to spend a couple of nights at his home, and there is to be a family outing. On the way, I am going to buy Arun his first laptop and mobile phone.

We go to the commercial hub of the city and there Arun quickly decides what he wants. Despite his shy, quiet ways he knows his own mind and is fully conversant with the latest technology. He is very grateful for these purchases – a small dream come true!

It is a twenty minute ride in Amosh's taxi to bring us to Rajesh's house. This is where I came a year ago when it was half built. Construction work had stopped through lack of funds and the family were living cooped up in one room downstairs, probably wondering when they would ever see the light of day. With some financial support from me, here now is the completed house, fully constructed, beautifully decorated, rising to four storeys and with magnificent views from the upper terrace over the surrounding countryside. It is quite easily the most beautiful and spacious house I have seen in Nepal. In fact, with its modern kitchen, its baths and showers, its smart curtains and artistic touches of paintwork, I can only describe it as luxurious.

Rajesh's wife, Soni, whom I remember as a model of style and elegance from last year, greets us at the entrance gates with her children, Sara and Basanta, beside her. I can't help reflecting how these two adopted children were removed from their humble rural background about six years ago and are now living in such luxury, so untypical of the average Nepalese family. It is an extraordinary contrast and I wonder if their natural parents have ever been invited to visit them in their new home. (I also can't help wondering how Rajesh manages to live in such a lifestyle on a pastor's salary.)

I am shown to my spacious room (complete with en-suite bath, shower and balcony) and later, over a cup of tea, I hand out my presents from England. For Soni, there is a brightly coloured silk shawl, and for the children, illustrated abridged copies of *Robinson Crusoe* (for Basanta), and *The Secret Garden* (for Sara). Over the weekend I will read some of each book to them out loud, then hopefully they will be able to carry on by themselves.

Before it gets too late, Rajesh and I, Arun and the children set off for an hour's walk in the warm afternoon sunshine. I remember these footpaths from last year and again enjoy the fresh clean air and tranquil atmosphere of this area. I notice that there are several detached houses not unlike Rajesh's and he tells me that many politicians and senior civil servants have their homes in this district, because, apart from the unpolluted environment, there is also a plentiful supply of good quality drinking water. He is pleased with the children's school but is thinking of moving them to one a little closer to home, because the present one is quite a long walk away.

In the evening I meet Soni's brother, Rashal, and his wife Sumetha. They are much younger than Soni, perhaps only about 20 and 18. Apparently, they ran away from their homes to get married a year ago but have since regretted it because they did not complete their education and so are unqualified for any job. I wonder if they may be able to help with Rajesh's orphanage project one day. They live close by, with Soni and Rashal's mother, Purna Devi, who I will be meeting tomorrow. Maybe she will also get involved, as well as Indra Maya. I can imagine Mama coming to live here when the five boys have left home. This house is big enough to accommodate plenty of people as well as the eight orphans Rajesh is planning to take in.

The family outing takes place the following day. We are going to hire a taxi and, on Sunil's recommendation, visit Namo Buddha, a Buddhist stupa high in the mountains, a couple of hours' drive east of Kathmandu. Rashal and Sumetha are going to look after the house, while the rest of us cram into the small taxi and set off on our pilgrimage.

I am given the front seat while Rajesh, Soni and Arun are squashed into the back with Sara and Basanta sitting on their knees. Outings like this are rare events in Nepal, so there is a feeling of excitement in the air. It has been laid on partly for me, as an honoured guest from England, partly for Arun, who is cheerfully poised with his new phone to record the proceedings, and partly for the children who have just completed their exams. The only problem is that nobody knows the way - least of all the taxi driver - and this is to prove a major handicap as the day unfolds.

By ten o'clock we are heading off into the countryside, driving through extensive fields of wheat and potato crops. The roads are becoming increasingly poor; in fact, they can scarcely be called roads at all, but are more like unmade tracks full of rocks and potholes. Sometimes our driving speed reduces to little more than a walking pace. The taxi driver occasionally asks the way. "Straight on, straight on!" the passers-by continually indicate, so on we go, through tiny remote villages and isolated communities, past forests and farms, until eventually we make out a distant mountain. There, right at the crest, we can dimly make out the hazy fluttering of Buddhist prayer flags. This must be our destination, but how will we ever get up there? The roads are not likely to improve, so we will probably have to get out and push.

At least we know we are heading in the right direction, so we begin the very slow ascent. The view opens up and we have plenty of time to admire some fine scenery as the little taxi struggles to convey us to the sacred site. Soon it comes to a complete halt. The road has turned to sand and we do indeed have to get out and push. Some passing cyclists stop to watch with interest. "How much further to go?" we ask. "Oh, it's only a short way now," they assure us and, looking up, the stupa is now clearly in sight, but it still looks a pretty steep ascent.

After five minutes of pushing, the road reappears and we are able to resume our painfully slow and bumpy ride. Basanta and Rajesh lighten the load by walking; soon Arun and I follow suit. By now it is 1:00 pm and very hot. I find it preferable to walk the remaining distance rather than endure the increasingly uncomfortable car ride. Eventually we all meet at the summit of the mountain, where there is a small car park, a few shops and houses, a café, half a dozen tourists and, up ahead, the gleaming white stupa of Namo Buddha. We feel a sense of achievement at the end of our pilgrimage.

Slightly shaken by the journey, we walk slowly through the little village to pay our respects at the shrine. It has been worth the visit, and we spend some time taking photos, enjoying the view and admiring the gifts on display in the local shops. Various garments are laid out to dry on the nearby bushes. In the street an argument breaks out between an old man armed with a stick and a frenzied old lady. There is a group of Polish girls who have made the journey here on foot. Their guide is conducting them to the one local restaurant, so we decide to follow suit. It has been a long morning and we are ready for lunch.

The restaurant has a curious arrangement: the seating area for patrons is on one side of the street, while the kitchens are on the other. So the meal demands a lot of scurrying across the road on the part of the waitress, while balancing large dishes of *dal bhat* and cooked vegetables on her outstretched arms. With about sixteen people to serve all at once, she manages extremely well. By 3:00 pm we have paid our bill and are ready to set off for home. I only hope our waitress will have a restful afternoon after all that running.

I am slightly dreading the homeward journey, but our taxi driver manages to find a different route and we do not have to face again the horrors of all those unmade roads. On the way, Rajesh leans over to me:

"Nick, the children are looking forward to the toys you promised them. Can we go to a department store on the way home?"

I had quite forgotten I had offered to buy them some toys, but I say, "Yes, let's go there."

It is almost a pleasure to be back in Kathmandu after the trials of the day and when we arrive at the shop, it is time to say farewell to our taxi driver. I thank him profusely for driving us there and, more importantly, for bringing us safely back again. He nods and shakes his head with great delight. I think he has also enjoyed his day out.

In the department store the children quickly find what they want: Sara opts for three cuddly toys (including the largest teddy bear I have ever seen), and Basanta picks out two remote-controlled cars. So all is happiness by the time we arrive home. The children have their toys, Arun has his photos to scroll through, and I have a shower, a cup of tea and a rest before supper. It has been an unusual but memorable day.

I have brought with me from England a donation of £100 from the Unitarian Christian Association towards the work of Rajesh's parish of Sangla. Next day, just before we set off for the Saturday morning service at his church, I hand this money to him.

"My Unitarian friends in England were touched by my story about Nepal and made this spontaneous collection for you at their last gathering. I told them it would be much appreciated."

"Thank you very much, Nick, and I will write them a letter of thanks later on. This money will be very useful for us, I promise you."

So we set off on the half-hour bike ride which Rajesh makes twice a week to his parish, once for the church service on Saturday and once in the middle of the week for house visits. There are six people at the service today: the parents of Sara and Basanta together with one of their other daughters, Simon the cobbler, and myself. I remember them all from my visit last year, as they do me, and we greet each other warmly. Rajesh reads from the Bible, leads prayers and worship, and we sing hymns to the accompaniment of Rajesh's guitar and Simon's drum.

I am asked to tell them a story, which I am happy to do while Rajesh translates. The moral of the story, which concerns a king and his visit to a wise old man, is threefold: the most important time is the present; the most important person is the person you are with; and the most important thing to do is to make that person happy. It is a simple story and I hope the message has come across clearly.

We arrive home for a late lunch, but soon it is time to depart. Arun and I say farewell to Soni, Sara and Basanta, thanking them for a lovely time. Arun has loved being with the children, but now can't wait to get home and set up his computer. Before that, however, we have two small commitments to fulfil.

The first is to call on Kaji, the guide on my first trip to Nepal, his wife and daughter. They live nearby and I have promised to visit them following my failure to do so last year. Within about twenty minutes we are outside their small shop and there is Kaji at the front door to greet us. It is so good to see him again after nearly three and a half years. He looks fit and well, stocky and upright.

"Hi, Nick. How are you?" He takes me firmly by the hand.

"I'm fine, thanks, Kaji. You're looking well."

Only Amita of his daughters is at home and seeing her again takes me back to our meeting at the International Guest House on my first day in Kathmandu. We are pleased to see each other, and I introduce her to Rajesh and Arun. We go inside to the room at the back where she installed the computer I bought for her. She says she is still very pleased with it.

I ask Arun to show her the pictures he has taken on his mobile. This forces him to talk rather more than he is accustomed to, but he seems happy to give Amita a running commentary on the weekend's events and it engages them in conversation for a while. I then ask Amita what she is doing nowadays.

"I have taken my final exams in English Literature and Journalism, but I have to wait until next year for my result."

"Next year? That seems a long time away."

"Yes, but that's nothing unusual here."

"Are you able to get any work meanwhile?"

"Not really. It's quite difficult. But I have managed two weeks as a trainee reporter."

"Did you enjoy it?"

"Yes, but it is tough work trying to get round the streets of Kathmandu as a reporter!" I can imagine it is.

I then tell Kaji about a plan I have been hatching for some time.

"I am hoping to come back to Kathmandu in October and take the five orphans on a trip to Pokhara. I am wondering if you would be our guide."

"Yes, that would be okay. I know Pokhara well."

So that is easily settled. We are given drinks and refreshments, and soon it is time to go. I shake hands with Kaji's wife on the way out and thank her for her hospitality. As we set off, I think more about my proposed trip and how nice it would be to take Mama as well, and perhaps her daughter Ira for female company.

We now take a microbus to Thamel where I want to buy some Yak wool shawls and socks. With Rajesh's help and bargaining skills I buy ten of each, some of which I will give away as presents, and the rest will go to Barbara Datson as raffle prizes for her charity.

Amosh now arrives with his taxi as if by magic and takes us the rest of the way home. Soon we are back at the orphanage and it is good to step inside the boys' simple home after all the spacious splendour of Rajesh's house. Anish and Sunil want to know about our weekend, so we tell them about our snail-like taxi drive along unspeakable roads the day before.

"And I believe it was your suggestion, Sunil, that we should go to this remote place high up on a distant mountain. You probably knew how bad the roads were going to be!"

When I tell him the name, Namo Buddha, he laughs out loud.

"That was not the place I suggested," he said. "Rajesh obviously did not hear me properly. You went to the wrong place!"

He thought this was a great joke – typical!

# Chapter Eighteen

A general strike has been announced for today, Sunday, which means that the streets of Kathmandu will be completely quiet as well as traffic-free, dust-free and pollution-free. It's almost unimaginable, and I can hardly wait to experience this oasis of calm in place of the usual din and chaos. Even as I set off on my early walk with Sunil, I can sense a quieter atmosphere in the neighbourhood, that Sunday-morning feeling which I can remember from long ago in England.

I love these walks with Sunil. He is such pleasant company: attentive without being fawning, jokey, thoughtful and keen to share his feelings. Today he tells me about his Marxist leanings and why he supports the general strike.

"Nepal is a backward country, and one reason we are held back is because we still have a caste system. My family is Kshatriya, which is the warrior caste. We are fighters, but also thinkers. The Marxists want to do away with the caste system completely because it is old-fashioned and too traditional. I support this, so I became a member of the student Maoist party and wave the red flag at the political rallies. But I may change one day. There are so many changes I would like to make to this country and one day I am hoping to go into politics.

"At Oxbridge College, I have been taking a course on Social Work and I have been influenced by reading about the history of charities in England. Do you know, uncle, the UK was the first country which teaches people to be civilised, to help each other, to sacrifice for others? This has given me so many ideas, and with my friends I have opened a club called "Our Dream". Its aim is to educate adults in rural areas to read and write, to bring them solar power, and teach them handicrafts and so on. Also we have a future plan to run a women's employment programme."

Being President of his college is giving him good experience at settling disputes and I tell him the meaning of such words as 'compromise', 'arbitration' and 'solidarity'. As we walk along the main road, now only occupied by a few armed militia maintaining the strike, we see a large billboard with pictures of Sydney Opera House and London's Tower Bridge. "Want to study abroad?" it says. " Ring us on…." Sunil puts the number into his phone – he will ring it sometime after I have left.

We are now in an area unfamiliar to me, and when we turn a corner, Sunil calls out to someone who is walking uphill away from us.

"Kiran!"

The man turns round, and starts walking in our direction.

"This is my older brother," Sunil tells me. I had quite forgotten Sunil had a brother besides Niran.

They talk for a little while, then Sunil turns to me and says, "He would like to give us tea. Is that all right?"

I say yes, so we set off together in a different direction. On the corner of a street is a three-storey building with a small shop on the ground floor. Kiran shows us up to his room on the first floor, then goes down to buy some milk from the shop. This room is his home, with single bed, Calorgas cooker, lots of children's cuddly toys, a few Christian pictures such as *The Last Supper*, and a tiny sitting area with TV and table. This is where Sunil and I sit.

"Kiran has a wife and little girl, but they live with his wife's mother most of the time."

I don't like to ask why, but guess that the mother-in-law is unwell and needs looking after.

Kiran returns with the milk and for the next twenty minutes sets about making our tea with the most elaborate preparations. The cups and saucers are washed, then washed again in a red plastic bowl before being dried to a fine polish. The milk is poured into a saucepan on the gas ring, tea and sugar added, and then carefully brought to the boil, all the time being stirred with a small plastic sieve. While this is going on, Sunil continues talking to me quietly.

"I am not really supposed to be here, because Indra Maya does not like me meeting my brother," he says.

"Why is that?"

"For some reason she is prejudiced against him, so you mustn't tell her we are here because I would get kicked out of the orphanage."

"What does he do for a living?"

"He is a stitcher, a tailor - and a very good one. But he has had a tough life without any of the good chances that Niran and I have had. When we were orphaned as young children, we got split up. Niran and I were taken into the home of a Christian man who looked after us, but Kiran was left behind in our village where he was forced to work as a houseboy. He was always being beaten and scolded. So he ran away to Kathmandu and somehow found his way to the house of a good man who taught him stitching. There was a big fashion centre in Kathmandu which needed one more worker, so Kiran went there and did a Masters degree in tailoring in his late teens. Niran and I then lost touch with him for six years.

"I next saw him when he turned up at Wasta Care Centre where Niran and I were then living. He had not been able to get regular work in Kathmandu after his degree so he borrowed money and went to Malaysia to get work. He had a little success but it was not very good, so he came back and had to pay back his loan. He then got married and soon they had a baby girl. Still there was no work, so he went to Dubai, but there was nothing there either, so he came back. It was then that he found Niran and me at Wasta Care Centre."

"How did Mama react when he turned up?"

"She said she could not help him. When he left, Niran and I were told not to see him. But we did, secretly, because we knew he was desperate and struggling with his wife and baby. We found out where he lived and would meet him in secret to give him our lunch money. But then his wife and child became sick and he had to try to earn a living. So he took out another loan, opened a small shop and spent what he earned on doctors' fees, but he could not pay the rent and had to give up the shop.

"Then Kiran met a man who needed a tailor. He employed Kiran and even gave him a shop to work in. My brother had a lot of customers and his life was improving. But soon this man's heart changed and one day he said, 'I must have my shop back.' So Kiran was back where he started. Eventually, he found work at a clothes shop which is like a factory, working from 10:00 in the morning to 7:00 at night. This is where he works now. He does not always receive his full wages and gets depressed about his situation. I keep encouraging him, and tell him just to hold on because life will improve and I will be able to help him later but it is very difficult for him. He says that he has no hope in life, and has even thought about committing suicide."

By now our tea is ready, and I sip it thoughtfully, while the two brothers speak quietly together. Kiran must be one of the gentlest, meekest men I have ever met. He is 26, I learn later; he does not look like Sunil at all but does remind me of Niran, with the same soft, round, fleshy face and mild eyes. I wonder if there is anything I can do to help, but cannot really bring myself to say anything just now. I will talk to Sunil later.

There is no work for Kiran to do today because of the strike, so he will just be staying at home. I ask him if he could make me a pair of white cotton pyjamas before I leave on Tuesday. That would be no problem, so he takes my measurements and asks me whether I want cord or elastic. "Cord, please," I say. Pockets? "Yes, pockets as well, if possible." Would we like some more tea? "Yes, please, that would be very nice."

While he prepares the tea, I am still wondering if I can help him in any way. Eventually I ask Sunil, "How much would it cost for Kiran to clear his debt and set himself up in a small shop?" The brothers have quite a long conference about this, then the answer comes back: about six lakhs, which is 600,000 rupees. One lakh would clear his debts, and five lakhs could set him up in business.

"Tell your brother I would like to help him," I say. (I had calculated this at about £4,500 – a sum which I could afford and which would totally transform this young man's prospects.)

They talk a bit more and then I am handed my second cup of tea.

"Kiran is very happy about this. He says he has never been given such a chance in his life. Look, he's got tears in his eyes. I have never seen him cry before."

Kiran cannot speak English, so we simply hold hands and look at each other for a few moments. I feel that, if nothing else comes from my trip, then at least pulling Kiran out of the abyss of debt and despair is worthwhile.

"Tell your brother I will promise to do this as soon as I am back home. You must send me your bank account details and I will send you the money."

Kiran says something in reply which Sunil translates for me:

"Thank you very much, Uncle. I will start looking for equipment and premises as soon as I can. I promise I won't let you down."

We part company for now, and I will return on Tuesday morning when my pyjamas will be ready. Sunil and I walk back along the main road, enjoying the effects of the general strike: total quietness, apart from the occasional small group of officials in the middle of the road pulling their weight to ensure that the strike is properly observed. One may walk or cycle, but that is about all. By 4:00 or 5:00 o'clock this afternoon, it will be officially over and normal life will resume.

Back home, after a late breakfast, I ask Anish and Sunil if they would like to go out and do some sketching. They are interested in the idea so I gather together the few art supplies I have brought with me, reckoning there will be enough for us all to have a go. Karjun has turned up this morning, so at about 10 o'clock we set off for a walk, in the hope of finding a good sketching spot. There is a real holiday feeling in the air and it is quite liberating to know that there is nothing else to do but enjoy ourselves. In fact, with only two days of my trip remaining, I feel I have done everything I came to do, so now I can just relax and take it easy in the company of the boys and their family. It is a good feeling.

After walking and climbing for an hour or so, we look down and see a possible site. Below is what looks like a Buddhist temple situated on the banks of a river. On the opposite bank, in the shade of some trees, is a shelter where we can sit to draw the scene. We scramble down the hill and find a stone wall to sit on. Yes, it is a good subject: a temple, a bridge over a river, hills behind, and a small hut with vegetation in the foreground. I hand out paper, pencils and felt-tip pens to everyone and we settle down for an hour or so. It is a welcome break, because the temperature is rising rapidly and here we can rest for a while in the shade of some overhanging trees. In the end, I work with Anish with some watercolours. I show him how to mix and lay washes, how to work from light to dark, how to simplify and adjust the scene in front of us into an attractive sketch. We become totally absorbed in this and I can't quite believe what is happening: here I am giving a lesson in watercolours to one of my godsons in Kathmandu. The result is nothing special, but Anish is thrilled with his first picture. He wants me to keep it as a memento of our expedition.

Sunil and Karjun have been doodling, meanwhile, and soon it is time to set off. We amble back through the community of houses and shops in the neighbourhood, returning for a late lunch of vegetable *momos*, specially prepared for me by Mama. Now that I am in drawing mood, I tell the boys that I would like to do their portraits. They are happy about this, so while they are sitting still working on their computers, I draw first Anish and then Sunil. Although I am really no artist, I find drawing the most absorbing of activities. The rest of the day seems to drift by in a relaxing pleasant way – talking, sketching, drinking tea, stopping occasionally to look at photos or discuss something. I feel very much at home in their company, but it is still amazing how my life's journey has brought me here.

# Chapter Nineteen

I am due to fly home on the evening of Tuesday 8th April, but today, Monday, is a better time for all the family to gather for a meal, speeches and farewells. This will not happen until about two or three o'clock, so I offer to take Anish, Sunil and their cousin Ashish to buy clothes for their new motorbikes – jackets, trousers, gloves and helmets. They are very excited about this and we set off on foot to catch a bus. A shopping expedition in Kathmandu is a long and exhausting process, what with density of traffic, the polluted atmosphere and then the problem of finding exactly what you want. Ultimately it is successful, but by 3 o'clock I announce that I am tired and please can we go back by taxi? They happily oblige and we are soon home, where many family members have gathered for the farewell party. I am particularly pleased to meet Ira, Mama's only daughter, and her two children: Indu who is 16 and hoping to become a nurse, and Rohit who is 14. Ira is a leader in a church called Chabel, where Amosh is the pastor; she is married to Dhan Krishna who runs a butcher's shop.

Eventually about twenty people are assembled in the living room of the orphanage, and the singing begins. Ashish plays guitar, Sunil is on the drum, and we sing some

Nepali hymns which are now becoming quite familiar to me. Everybody knows the words by heart and I join in as best I can, either making up words of my own or humming along with everyone else. At the end of this session, Rajesh announces that I will be departing the following day and invites everyone to say prayers for my safe journey home, which they do quite vociferously.

Now Mama wants to say a few words, which Rajesh translates so that I can understand. She thanks me for all the support I have been giving her family, and shows me a picture of a man on a riverbank pulling another man out of the river.

"This is you pulling us out of the deep waters of trouble," she says.

She hopes I have been happy here and says that I will always be welcome to stay with them in future.

Rajesh now turns to me and asks if I would like to say something in reply. I say I would and he translates the following short speech which I make:

"First of all, I would like to say thank you for the very warm welcome you have given me on my second time staying with you. I had a particular reason for coming this year which was to get to know the other three boys, Anish, Sunil and Arun. I want to support them as well so I needed to discuss their futures.

"Last year, I spent most of my time with Ramesh, as well as Niran. With those two away at college, I have spent my time with the other three, and I have enjoyed getting to know them better.

"One morning last week, Anish took me on a walk and we visited the first and second homes of Wasta Care Centre. This was very interesting for me because I am gradually forming a clear picture in my mind of the early history of your family, the hard times of suffering as well as the loving bonds of friendship which have been growing among you over the years. It is truly a wonderful story, which Anish helped me to understand a little more. I am pleased with his progress in the hotel management course, because I think he has found a line of work which really suits him. I hope one day he will open a restaurant in Nepal, and if so I promise to be one of his first customers!

"I had a wonderful trip away with Sunil and Rajesh to Chitwan and Lumbini. Sunil looked after me very well during this time, attending to all my needs and chatting to me all the time. He pulled my leg quite a lot – and also liked to pull my hair, because he thinks I should have the same style as him, with my hair sticking up like a cockerel. When we came home, he wanted to show me some of his pictures. And I noticed that most of them were pictures of himself! So I think that he quite likes the look of himself and admires his own appearance. He tells me he enjoys making speeches and being the centre of attention, and I think one day he may enter the political arena. I can imagine him being quite a fierce politician and I certainly would not like to oppose him. Sunil has been a great companion to me, with all our walks and outings, and I have enjoyed getting to know him better.

"I have also enjoyed Arun's company, and it was great to have him with me when we stayed at Rajesh's house. He would tell me to take plenty of rest, he would keep all the zips of my bag done up, and he even told me to wear a vest if he thought I would catch cold. He does have a very soft voice, so I have to lean close to hear him, but this does not matter; and at least we have been talking to each other more this year. I think he enjoyed the day out when we had to push the taxi up the hill. He took pictures every step of the way on his new phone, as this was his first ever trip outside Kathmandu. My home telephone number was the first number he put into his contacts list, which is a great honour. I only hope that, if he rings me up one day, I will be able to hear the quiet, whispering voice at the other end.

"I have also seen more of their cousins Ashish and Abhishek this year, and have been enjoying their company too. I will always think of Ashish as a writer and singer of songs. He tells me some of these are worship songs and others are love songs. At the moment, I cannot quite tell the difference between them. They all strike me as heartfelt and very personal, and I loved hearing him play and sing on the guitar. He has a great gift and I hope he continues to develop his musical talent throughout his life.

"The other day I was shown a short home video of Abhi being pursued by a camera, running away down a hillside path looking both scared and amused. He was the star of this film, because he was the only person in it! I took this as a sign that one day he might become an actor and a great movie star – but he is still only 14 and probably has lots of other ideas of what he wants to do. I know he is good at drawing, so I have given him a sketchpad and some fine-line drawing pens to draw pictures of you all. This is my commission to him. As some of you know, I am writing a book about my experiences in Nepal, and all of you will feature in the story. Abhi will, I hope, provide the illustrations and portraits, and then the book will be a permanent record of how I came to be a part of your family.

"So my main role is to act as godfather to these boys, to support them in their education and to help and guide them should they ever wish to share their concerns with me. I have loved getting to know them better, and I will always be interested in their progress through life. At the moment, I have this idea of coming again in October, when I would like to take all the boys and Mama to Pokhara for a short holiday. I have already asked Kaji if he would be our leader and guide and he has said he would, so I hope this can be arranged."

There is a little clapping when I finish so it seems to have gone down well and I am of course very grateful to Rajesh for translating. His English is extraordinarily good for a man who has had so little formal education. Nobody seems to want to leave now, maybe because they don't all get together very frequently, so they continue to chat for a while longer.

But there is one thing I still want to do before I go home. I whisper to Sunil who is sitting next to me:

"Can we go for a haircut and a head massage?"

"Yes, of course. I know a really good barber. I always go there. Come with me," and we quietly slip away from the family gathering. Ashish decides he wants a haircut as well, so we set off together.

We have quite a long walk ahead of us, but I can see why Sunil takes the trouble to come here. The barber, we learn, has been working at his trade for twenty-five years, so he really knows his business and his customers come from far and wide. He starts work at 7 o'clock in the morning and apart from an hour off at lunchtime he works through until 9:00 at night. In fact, it is in the evening when most of his regular customers come.

We are each given forty minutes of the most attentive and enjoyable hairdressing experience imaginable. After the dry cut comes the head massage, then a neck and shoulder rub, then back and vertebrae. Some of the time we lie head-down on a towel; other times he pulls us upright and pulls our shoulders back with the full weight of his arms. The final flourish comes with a flat, electric vibrating machine, which he manoeuvres over our head, shoulders, back and neck. It is the most wonderful haircut I have ever had. Watching Ashish going into raptures compels me to photograph his expression of delight for posterity. Afterwards, I teach Ashish a new English word for his vocabulary: 'ecstasy'!

We return home looking smart, as well as feeling revitalised and refreshed. Now the boys want to show off their new motorbike outfits to all the family. They try out each other's jackets and helmets as well and, of course, a great many photographs are taken. I promise I will send the money for the bikes as soon as I get home. By now, most people have left and the house has quietened down. While Ashish is watching TV, I decide to draw his portrait: Abhishek sits beside me watching, while I try to explain what little I know about how to draw someone. But I am aware that he probably has much more talent than I have, and I am sure he will produce some good illustrations for my book.

The next day marks the end of my trip as I will be flying home in the evening. So at 6:15 am there is a knock at my door and Sunil is ready to take me for my final pre-breakfast walk. Knowing that I have two long flights ahead of me, he gives me nearly two hours' exercise before we finally end up at his brother Kiran's home. This time his wife Sumetha and little girl Miriam are there, so the room with all its toys and children's books looks more complete. Miriam has long, jet black hair and wide bright eyes: to me she looks like a gypsy girl, as there is a sort of wildness to her. While she is being made ready for school, she sings some songs – and I soon recognise them as old-fashioned nursery rhymes which she is singing in English. Extraordinary how far the tentacles of our cultural traditions have spread! Today is important because the results of her recent exams are coming out. By eight o'clock she is ready, so Sumetha sets off with her, leaving us behind to have some tea.

Kiran and Sunil chat for some time and then Sunil turns to me:

"Our brother Niran in Delhi spoke to Kiran last night and said he was very lucky to have this opportunity to clear his debts and start a new life in a tailor's shop. He said that if he wasted this chance, he (Niran) would never call him 'brother' again."

Apparently Kiran has already found where he can buy a sewing machine and the other equipment he needs and he knows where he wants to set up his shop. He has made my pyjamas but they need ironing (he does not have an iron here) so he will give them to me later.

After about half an hour, mom and daughter return. Great news! Miriam has come second in her class of forty pupils. Sunil has already told me she is a bright girl, but to do this well is really quite an achievement. She shows me her exam booklet which, at over thirty pages, I think is pretty long for a little girl. From the quality of her answers, handwriting and drawing I can see why she has done so well.

We say goodbye and set off for home, where my usual breakfast is laid out on the table, this time with the addition of a croissant.

"Uncle, I have just heard from Kiran. Let's go and collect your pyjamas," Sunil says quietly. "Nobody must know, so we will just go out secretly."

This secrecy seems such a shame to me, but I have to go along with it. We walk up the road and there, at the busy intersection, is Kiran holding a small package. He smiles modestly and then has a few words with Sunil.

"He would like you to accept this as a small present," says Sunil.

"Please tell him that I am very pleased and I will always think of him when I wear them."

We shake hands, and I pat him on the shoulder wishing him the best of luck. We walk away but after a few steps turn back and wave to each other. I think this has been quite a special encounter for us both.

There is not much left to do now except pack and say our farewells. Two of the boys want a private word with me, so it is difficult to make much progress. First Arun comes into my room, wanting to thank me again for his presents and saying how much he has enjoyed my being here.

"This has been a wonderful time for me, Uncle."

Then Anish comes in. He wants to talk about his hotel management course, which he is not totally happy about. He finds it too theoretical and not practical enough. He wants to do a training course combined with working at a hotel at the same time.

"Do some research, and send me an email, telling me what you find out. If it really does seem a good idea, then I will be quite happy for you to make a change. But keep me informed."

The car is now being packed up. Karjun will drive me, Rajesh, Anish and Sunil to the airport, so I have to say goodbye to Mama, Arun, Ashish and Abhishek. There is a slight feeling of sadness now, but we smile bravely as we embrace.

"Take care of yourselves," I say, "and I will look forward to being with you again, hopefully in a few months' time."

We wave as the car pulls away and now drive mostly in silence to the airport. It doesn't seem long since I was being picked up here – well, it has only been eleven days.

The final farewell is a bit more difficult and Anish cannot quite manage it as his eyes fill up with tears. I simply say,

"Thank you all so much for giving me a wonderful time. Whatever happens now, we will always have this holiday to remember."

I turn round and join the queue for departure. Looking back once, I wave to them again and then we go our separate ways.

✸ ✸ ✸ ✸ ✸

# Chapter Twenty

I haven't related yet what became of Santosh Karki following my talk with Lama Kondan. While I was staying at Rajesh's house, I asked if he could ring Santosh and arrange for him to come and have a chat. I was still feeling pretty upset by his deception, and asked Rajesh to talk to him first, repeating what Lama Kondan had told us. Santosh duly arrived, looking smartly dressed and carrying his new laptop, which he wanted to show me. Rajesh spoke to him in Nepalese and Santosh remained silent throughout. If he had any feelings of shame at what Rajesh said, he did not display them. He sat still and upright, with a calm expression on his face.

When he finished speaking to Santosh, Rajesh turned to me and said, "I have been telling him what Lama Kondan had told us, that he has not been totally honest, because he has already been receiving financial support from someone, which he did not tell you about, and now he is taking advantage of your generosity to receive more money. I have told him that the medical course is very expensive and that he must try to understand this and not set his sights so high. He can look for a good job – he is a bright boy, and we will always continue to support him in whatever he chooses to do."

I immediately felt there was something a bit inconsistent in this line of reasoning, since the family was happy to let me pay for Ramesh's education as a doctor, but because Santosh had not been entirely straight with me, this boy's future hopes were being utterly dashed. So I interrupted Rajesh:

"Let me have a word with Santosh now."

"Of course," said Rajesh, so Santosh followed me into my room. We sat down and I said, " I have not made up my mind whether or not to support you. But you must appreciate how I felt when Lama Kondan told us those things."

Without looking crestfallen or abashed he simply said, "I am sorry. I should have told you, but the fact is that when I was in Grade 6, a man from Hong Kong came to visit Triple Gem School and for some reason he took an interest in me. He thought I looked a bright boy and said he would support me. For two years he sent me money and then all of a sudden the money stopped coming. So I emailed him from time to time but I have not received any more money from him. I believe I have sent him four emails but I have not heard anything from him for twelve months. So really, I am not receiving any financial help now, but I do understand that I should have told you this."

There was a period of silence while I took this in and thought for a moment. I then said, "What were you planning to do after completing your final exams at school this year?"

Again, he did not look self-pitiful, but replied almost cheerfully, "I thought I would just have to get a job somewhere."

What a waste! Because of a system which denies higher education to all but those who can pay for it, this bright boy was likely to be consigned to the level of the average working man of Kathmandu with no prospect of anything better.

"What made you want to become a doctor in the first place?" I asked him.

"I had a sister who died of a snake bite. In my village we could not get medical help in time to save her. I was only twelve and I was so upset by this. She was walking through a field in the evening, with her older sister beside her when it happened. When I heard the news I ran through the fields with bare feet, but I was too late. I was so upset I felt like that I had fallen from the top of a mountain. The last moment I saw my sister's eye she was asking for life, but everybody was helpless. I go to the same field when I am home and looking at that field I feel like my sister is still working there, but in reality my eyes are filled with tears. I know that wherever she is, she is happy and safe - and I knew then that I really wanted to be a doctor, more than anything else."

I was very moved to hear this story and also at the boy's sincerity. I couldn't say anything for a few moments, but then I said,

"How are you getting on in your studies now? Do you think you will do well in your final exams?"

"Oh yes, I will do fine," he said with utter confidence. "I find them quite easy."

My mind was made up.

"As far as I am concerned, Santosh, what has happened in the past is over and done with. If you manage to get a place at a Medical College, I will sponsor you to be a doctor and give you the financial support you need. You seem bright and committed, so I feel you should be given a chance."

A look of pleasure and relief spread over his face, and he simply said, "Thank you very much. I promise you will always be able to trust me."

"You do not need to tell Rajesh or anyone else in his family. I believe you deserve the opportunity to train as a doctor, and I am happy to help you."

We shook hands, and I walked with him to the main road where he could catch a minibus to take him home.

"So are you happy now?"

"Very happy, Uncle."

"All I ask of you is that you keep in touch by email just as much as you can, however short the message. Always let me know what you need or want, and I will be there to help you. We obviously won't ever be seeing much of each other, but maybe one day you will come and see me in England."

"Oh yes, Uncle, I would very much like to do that."

We parted company and I walked back to Rajesh's house. Over the next few days while I was in Kathmandu, I exchanged emails with Santosh. The last one I received on my final day was not very clear to me and I asked Sunil (who I had taken into my confidence about my support for Santosh) to give him a ring to confirm the details of it. In fact, they had quite a long chat, and when they'd finished Sunil told me, "I like Santosh, Uncle, and I will be ringing him from time to time. I feel close to him as he is the same age as me, so we are already good friends. Leave it to me, Uncle, I will be keeping in touch with him, letting you know how he is getting on."

"Just like your brother, the tailor?" I asked.

"Just like him, too. Trust me, I'll be looking after them."

I was very pleased about this. Somehow Kiran and Santosh were especially important to me and had given this trip an unexpected significance. As I pondered on my journey home over everything that had happened, I reflected that they seemed like outcasts who were now being given a chance to find their way in the world. Apart from getting to know Anish, Sunil and Arun so much better, this gave me as much satisfaction as anything: to be giving a helping hand to the mysterious boy from Facebook and the penniless tailor of Kathmandu.

# Chapter Twenty-One

I wanted to fulfil my promise of returning to Nepal in October that year to give Mama and the boys a short holiday in Pokhara, where we would walk and talk, enjoy the mountain scenery and have Kaji as our guide. I wanted to include their cousins Ashish and Abhishek, and maybe Mama's daughter Ira as well. Finally, I hoped that Santosh might join us, so that he would become integrated and accepted by the family, and make friends with the boys. But before all this was to happen, I was stopped in my tracks and had to rethink what I was doing.

I had always heard that foreign aid is a two-edged sword and sure enough, my lavish spending on motorbikes, new clothes, laptops and mobile phones in March had had unfortunate repercussions. The quiet harmony of Wasta Care Centre was being disrupted by angry voices, jealousy and disagreements, as the boys (Sunil and Anish in particular) were now experiencing a measure of freedom and independence which had hitherto been beyond their reach. It would appear that I had in effect spoiled them and they were now over-reaching themselves. One day, I received a long email from Mama (written by Rajesh), complaining that they were being rude to her, sneering at her illiteracy and no longer grateful for the love and care which she had lavished on them for so long. She was so upset by their behaviour that it was making her quite ill.

In their turn, Anish, Sunil and Arun complained to me that she was making a misery of their lives, scolding them needlessly and at times becoming almost hysterical with rage. It was clearly an unhappy atmosphere which may quite possibly have been my fault. In trying to give the boys something of the lifestyle which my parents had given me at their age, perhaps I had done more harm than good. So I was feeling a little uneasy as October approached, and hoped that the presence of Ramesh and Niran at Wasta Care Centre would help to smooth some of the rough edges.

Arriving at Wasta Care Centre on 7 October 2013, I am greeted as warmly as ever, particularly by the five boys. But after a day or two it becomes clear to me that there is an atmosphere of unease and tension in the air. However I really don't believe that my gifts of money and motorbikes are the problem – in fact, these latter have been proving very useful, providing transportation for all members of the family. Something much deeper is at issue here. Mama continues to complain to me about the ingratitude and rudeness of the boys, while they in their turn speak of her constantly nagging manner and Rajesh's controlling demands on them. The cracks are beginning to show and I try to analyse why.

The first thing I realise is that this is not a normal family set-up. These boys are orphans and so, to a greater or lesser degree, have been psychologically damaged from an early age. And although they have been looked after for about fifteen years, this is not a normal family home and Indra Maya is not their mother. So despite all the very best intentions, there is something unnatural and, I feel, almost tyrannical about Wasta Care Centre. I now feel a peculiar sense of oppression in the home, which the boys are powerless to free themselves from.

I gather from a conversation with Rajesh that the Norwegian Church was only ever going to support the boys until the end of their school days, at which point they would be expected to go out into the wider world and earn their living. This would bring an end to Wasta Care Centre, as well as to the income Rajesh and Indra Maya receive for taking care of the boys.

One evening, about two days after my arrival, I have a long talk with Ramesh about all this. He has been deeply concerned about the tensions he had felt during an earlier visit, only to be accused by Rajesh of interfering when he tried to offer suggestions for improving matters. On return to college, he wracked his brains trying to find a solution to this ongoing problem, but nothing ever presented itself.

So now I offer my radical proposal which has only come into my mind while we have been talking. I point out to him that he and the other orphans are under no legal obligation to stay at Waste Care Centre, so the only way in which they are not independent is financial. My suggestion is that I provide them with an alternative home in Kathmandu. Short of actually adopting them, I will take on the responsibility of supporting them until such time as they are able to fend for themselves. Whatever course of education or training they follow over the next few years, they will need a home to live in, and I will undertake to provide this for them.

I think it is with a growing sense of relief that Ramesh listens to this suggestion. He nods occasionally, and when I have finished he says quietly,

"Yes, I think this will be a good solution to our problems. Really, Uncle, we are entirely dependent on you for our futures, and we are in your hands for whatever you can do for us."

We talk for a while longer as we both begin to realise the full implications of this proposal: a completely new start to their lives, free from the tyranny of Rajesh and Mama and the oppressive atmosphere they have been living under for so long. I tell him I will speak to Rajesh at the first opportunity, hoping that there will not be any unforeseen difficulties.

"Thank you, Uncle. I will not say anything to the other boys yet, but I think I already see some hope ahead for us. I truly believe this will be a solution to our problem."

I have the opportunity to talk with Rajesh the next day. When I have outlined my ideas, he replies,

"We have always been very grateful for what you have done for the boys, Nick, but you must realise that Wasta Care Centre has been giving Mama and me a livelihood for the past fifteen years. When the Norwegian Church set up the orphanage, it was on the understanding that they would pay the rent of the building and give me and Indra Maya an income to bring up the boys until the end of their school days, when they should get a job. So if you do this, and the Care Centre closes, we will lose our income."

"You must have known that your income would come to an end one day. But I suppose the Norwegian Church never offered you a pension?"

"No, there was no pension arrangement."

So clearly it is not in his or his mother's interests to allow the boys their freedom: the longer they are under this roof, the longer he and Mama continue to receive income and she has a home. His objection to closing the orphanage is understandable bearing in mind Mama's impoverished circumstances but I am not going to be deterred.

"All right, Rajesh, I understand what you are saying, but still the boys are under no legal obligation to stay here. They are not happy in this home any longer, and I want to offer them something new. I am prepared to compensate you and Mama, because I appreciate your circumstances, but you already have a home of your own with plenty of room to take in Mama. Also, she can stay in the old family home with Karjun's family in Pulchowk, as well as Radheshyam's new house. This will give her a much more varied lifestyle than she has at present, and I'm sure in time she will be made welcome in the boys' new home as well."

In principle, Rajesh agrees with all this, only adding that he is under a debt of £6000 for his new house and would I pay him this amount? I point out to him that I have already given him a lot of money for his house and family and now that I am undertaking to provide a home for the boys, I am not prepared to give him that much. I offer to give Mama £5000 and him £3500, and he can ask her to help him pay off the rest of his loan if he wishes. He accepts this gratefully, and I feel I have been quite fair with him.

It is now time to impart all this to Mama, but very gently and tactfully because it will come as a big shock to her. The three of us get together later in the day and she listens passively while Rajesh spells out what I have been proposing. At first she seems to accept it all quite calmly – her lips are tightly drawn but she makes small nods of assent. Rajesh speaks for a long time and as I watch her face, I know that tears are forming. She takes her shawl and starts to dab her eyes. I stand up beside her chair and when I put my arms round her shoulders, she buries her head in her hands and gives way to weeping. I imagine it is like losing a son or a daughter when they leave home to be married. After raising these boys for fifteen years she will now be losing them all at a single blow. I suddenly feel awful – what right have I to be interfering with her life and causing her such distress? We remain together like this for a few minutes.

Soon she is able to control herself and speak to her son. I wait patiently until she has finished, but her speech is faltering as she fights back the tears and takes short gasps of breath. Rajesh interprets for me:

"Mama thanks you for the love you have shown the boys in paying for their education. She also thanks you for offering to find a new place for them to live. Now she wants to start looking after poor widows at her church, and she will live sometimes in the family home at Pulchowk with Karjun's family, or with me or with Radheshyam in Buddhanilkantha and sometimes with the boys in their new home. She thanks you for the money, which she will save and keep for her old age."

I hold hands with Mama and thank her for being so understanding. I am sorry for the sadness she is feeling, and realise how difficult it is to accept such a big change in her life. But I think in the end it might lead to a happier time for her, knowing that she no longer has the difficult responsibility of looking after the orphans. She seems to accept this with a good grace.

That evening I have a meeting with the boys. Ramesh knows the gist of what I am going to say, but the others have no idea.

"It has been quite clear from your emails over the past few months that you are no longer very happy living here. I do not think this is anyone's particular fault; you have simply outgrown Wasta Care Centre and it is time to make a new start. I have spoken to Rajesh and Mama about this, saying that I have offered to take care of you for the next part of your lives. With your help, we will find a new home for you in Kathmandu where you can live together freely and independently. I am not quite sure how I will manage it, but I cannot think of any other solution to the problem. What do you think of this idea? Do you have anything to say about it?"

I look round the room and they are staring back at me with stunned expressions, not knowing quite what to say or how to react. They remind me of caged birds when the door to freedom is opened. They are too wary to take the first steps outside their little home. Fortunately, Ramesh takes over from me, talking rapidly and at length in their own language. He is being upbeat and positive, but still is eliciting little if any response. Sunil looks around, glancing from face to face, but the other three are giving nothing away.

Eventually Ramesh stops talking, so I pick up the threads.

"Mama agrees that this is a good idea but at the same time she is upset about losing you after such a long time, so you must be nice to her and comfort her. I think she will find the change quite difficult after all these years."

The boys turn and look at each other, but still remain completely silent. As I have them all together, I decide to talk about something else.

"I now want to talk about you and your futures. To start with, I am very happy about Ramesh at Lumbini Medical College. He enjoys his life there; he is passing his exams well and hopes to qualify in 2015. And Niran still enjoys studying accountancy in Delhi, even though the course is very tough and he has to pass all the exams together, otherwise he has to resit them all. But he is determined this is what he wants to do, so I admire him for that and wish him good luck."

I now turn to the two brothers, Anish and Arun, who are much less confident than the others. Arun has a serious speech impediment making it difficult to hear what he is saying and Anish can be painfully shy and introverted, particularly in unfamiliar situations.

"Arun has just started at Oxbridge College and seems to be settling in happily there. He has asked if he can do extra tuition after school because he finds the pace a bit quick for him at times. This sounds like a good idea.

"Anish has not been happy at his Hotel and Catering College, but has now started a new 4 year course at a different college so I hope this works out well for him. His dream is to open a restaurant one day, and I think this is a great ambition to have.

"Sunil has recently started a Bachelor degree in Social Work (BSW) at Oxbridge College, and he has many plans to improve the welfare of the poorest people in Nepal.

"So really I am very pleased with all your aims and ambitions and am happy to continue supporting you."

I then turn the subject back to the matter of finding a house. I think we should start looking for a place while I am still here. Sunil suggests that they ask Prakash, their landlord upstairs, to give them some help and advice. He has always been on very good terms with the boys and, since his job is in real estate, he would get them off to a good start.

# Chapter Twenty-Two

So the next day, Ramesh, Sunil, their cousin Ashish and I are upstairs where Prakash and his wife Yamuna are making us tea in their living room. Once I have explained why we are there, Prakash speaks up (he addresses me as 'Uncle' which I find quite endearing):

"Well, Uncle, I have always loved the boys so much, I would do anything to help them. I never had brothers, so they are my brothers. I am sure I can help you, and then I will keep an eye on everything afterwards. Do you want to rent or buy a home?"

"If possible I would prefer to buy a place for them but I have no idea what cost is involved…"

And so we enter into a discussion of all the important details, which ends with an arrangement to begin house-hunting the next day. There are a few priority considerations, the first of which is to find a location outside the Ring Road where prices are cheaper. Secondly, the house must be on reasonably high ground where there is less danger of flooding and more chance of a good Internet connection. The third comes from their cousin Ashish: the boys must not be any further away than they are now, in fact much closer if possible. The bond between the five boys and Ashish and

Abhishek is so strong that to live too far apart is unthinkable. So this for Ashish is a top priority.

On Saturday we set off in heavy rain in a convoy of four motorbikes, headed by Prakash and Niran, on a house-hunting expedition. This really is the greatest fun, to be roaring along the narrow bumpy streets of an area known as Nakhipot, stopping on four occasions to view the houses which Prakash has to offer. The first is brand new, very stylish and beautiful but much too big; the second is locked and the agent involved is too busy playing cards to come out with the key; the third is only six years old but looks a bit shop-soiled. But the fourth looks promising. It is only one year old, has two good-sized bedrooms, a sitting room, kitchen and shower, while upstairs on the flat roof there is a traditional local feature which the boys find very appealing: strong pillars provide the basis for building a first and then a second floor which would be a good investment for the future. The location is a little further out than they would wish but this means the area is not too built-up. There are fields surrounding the property and views of the distant hills. Already it seems that the idea of the boys having a home of their own is becoming a possibility.

The next day is Sunday and during the morning there is a phone call from Kiran the tailor, who wishes to make me a pair of jeans. I am excited to be seeing his shop, so straightaway Sunil takes me there on the back of his motorbike. On the way, he buys me some stylish sunglasses to keep the dust out of my eyes. We park at a colourful and crowded open-air shopping area and from there walk into the Machindra Minimarket Shopping Centre.

"Kiran's shop is on the first floor," Sunil tells me. "Do you see above your head the sign I made to attract customers?"

I look up and there is a picture of two young women and one young man from the waist down, wearing jeans.

Between them is an image of the Christian cross, with the inscription "The Light Jeans Tailor". Underneath in Nepalese it says, "All kinds of clothes are mended here."

We go upstairs and beside the walkway I can see Kiran in his small shop, also his assistant and four sewing machines. With a tape dangling round his neck, Kiran is measuring and cutting a customer's jeans, I presume to shorten them. The customer and his friend are sitting on stools waiting for the work to be done, and after a brief nod and smile in my direction, Kiran continues swiftly with his work. He knows exactly what he is doing and it is wonderful to see him working with such sureness and confidence. There is another pale thin lad hanging around here too, but he doesn't seem to be involved in the business. I ask Sunil a few questions about how Kiran is getting on.

"Uncle, he is doing well. Many of the shopkeepers on the ground floor know him now, and they refer their customers to him for alterations. Kiran is popular and well-mannered, and he does good work for not much money. Because this is a new business, he cannot charge much so he is attracting many customers. He works long hours, from seven in the morning until nine at night, only taking off Saturday morning to go to church."

A few more people have arrived by now, including Ramesh and Ashish. Some seem to be here just to have a chat with Kiran and pass the time of day; others give him jeans or shirts to alter.

"Uncle, meet my brother, Raj", says Ramesh. This turns out to be the pale thin lad, but when we shake hands I begin to see the family likeness. He has the look of a lost soul, so once I find out that he can understand English, I chatter away to him quite freely. I gather that he missed out on his education when he was young, so now he is at school he is struggling to keep up. He lives with his uncle nearby, but the uncle no longer wants him in his house so this is a problem. He often comes to the shop because Kiran is warm-hearted and friendly and chats to Raj while he is working.

Now it is my turn to be measured for my pair of jeans. The best quality material is taken down from the shelves, and the two brothers, Sunil and Kiran, take a look at me.

"Uncle, we think you should have a slight bellbottom cut." I am a little dubious but they insist. "This style is popular here now." Oh well, I am sure they know best.

This is a one-hour tailoring service so after being measured, I go with Ramesh, Sunil and Ashish for tea at a nearby café, then take a walk through the market stalls. It is the start of Dashain, the main Hindu festival, and the streets are thronged with even bigger crowds than usual as people make their last-minute purchases before the shops close down for ten days.. Tradesmen are selling brightly coloured balloons, kites and toys from bicycles, whilst nearby the market stalls are piled high with rugs, shawls, clothes as well as vegetables and fruit. It is a noisy, crowded and colourful scene.

We return to Kiran's shop for my fitting. The jeans fit perfectly and he gives me a warm brown jacket as well. Apparently with my new sunglasses I am worth "clicking", as they say here for taking a photo, so I am clicked several times. I give Kiran a big hug of gratitude, before placing orders for ten pairs of pyjamas from customers in England. I will pick these up from him when I return next year. With Sunil as his accountant, I believe this business has a real chance of success.

As godfather to the boys, I want to offer them something more than just financial support. Perhaps I can share with them some readings which are not specifically religious. I feel they have had enough of this from Mama and Rajesh. Before I left England, I came across a self-help book by Deepak Chopra called *"Super Brain – Unleash the Explosive Power of Your Mind"*. I decide to read a few passages with them from the chapter entitled *"Finding Your Power"*.

So the next morning we are all gathered in the sitting room at Wasta Care Centre at 10:00 including Ashish, Abhishek and Santosh (whom I haven't yet seen on this trip). The chapter covers such topics as high self-esteem, self-confidence, the importance of an optimistic outlook and how to overcome obstacles in life. Occasionally I stop and ask them the significance of a sentence, or what such-and-such an idea means to them. The readings expand on what I was hinting at to the boys yesterday about seeking independence and fulfilment in their lives. Their response is fairly mute. Looking round the room, it strikes me that Ashish, Abhishek and Santosh have high self-esteem and self-confidence, in comparison to the orphans who look somehow oppressed and fearful. I am sure this will begin to change once they have a place of their own and are free of the past.

When we finish, Santosh wants to take me back to his room to give me some presents, and show me all the information about his medical college in Bangladesh. He did not pass the entrance exam well enough to gain a place at a college in Nepal, but he has been accepted by Jalalabad Ragib-Rabeya College which he is very pleased about. He has borrowed a friend's motorbike so he leads the way, with me following on the back of Sunil's bike. He sets a cracking pace – these two boys, who have become such good friends, clearly enjoy speeding round the streets, taking different routes and overtaking each other. I feel a bit shaken up when we arrive and have completely lost my bearings so have no idea where I am in Kathmandu.

Santosh's room is above a small hardware shop. It was totally bare when he took it on, so he set himself up with a bed, a chair, a couple of tables for books and a pair of thin curtains. I have to admire the self-possession of this young man who is almost alone in the world, but is taking charge of his life in a resourceful and intelligent way. He has no cooking facilities, but sometimes his landlord upstairs invites him for a meal. Otherwise he eats out at cheap restaurants.

He pops downstairs to the shop for cold drinks and when he comes back he says,

"Now would you like your presents, Uncle? Look, I have wrapped them for you."

Inside one box is a black Nepalese hat and a long thick woollen scarf, while in the other is a model of the famous Buddhist stupa at Bodnath, which I visited with Kaji on my first trip to Nepal. I am delighted with the gifts and tell him about my Buddhist friend, Barbara Datson, who goes for teachings to the White Monastery, near the Bodnath stupa. I remind Santosh that it is because of Barbara's charity that we two first met at Triple Gem School.

Now I give him my present, a silver and white wrist watch with a metal strap. His face lights up when he sees it.

"Oh, oh, Uncle, it is so beautiful!" Sunil adjusts the strap for him and soon it is sitting comfortably on his wrist. "Thank you so much. I will always treasure this."

He now shows me all the literature about Jalalabad Ragib-Rabeya Medical College. He will be flying out there early in January with twenty-five other Nepali students and they will be housed together.

"It has a reputation for very tough training at this college," he tells me, "much tougher than the medical colleges in Nepal. You will come and visit me, won't you, Uncle?"

I promise him I will make a special trip to come next year and see how he is getting on. Santosh has a bright personality and exudes confidence – I feel sure he will do well.

By the time Sunil and I arrive home, the others have convinced themselves that house number four is for them. I feel they should look around a bit more first, but maybe some practice at negotiation would be a good idea while I am still here. So we make arrangements to go and see Prakash later that evening to discuss terms. It turns out to be the most curious house sale negotiation I have ever known.

We go upstairs at about 8:00 in the evening to Prakash's bedroom, where he is sitting on the bed in his vest. His wife is sitting in the bed covered in a blanket watching TV. We are invited to sit on the floor and after a short friendly chat, the phone rings and Prakash hands the phone to Ramesh. It is the house agent with whom we will have to negotiate directly. The volume is on loud, so we are all able to listen in to the conversation, though of course I have no idea what is being said.

After a few minutes their conversation comes to an end, and then Ramesh and Sunil start talking with Prakash. Occasionally Anish chimes in and then to my surprise, Prakash's wife gets up from the bed and starts to become quite vocal. At least two conversations are going on now, but it seems very friendly and even quite jokey at times. Perhaps they are simply talking about the weather – it has been an unseasonably wet day, after all.

Prakash's wife then goes over to the wardrobe from where she produces a blue plastic bag full of documents. Perhaps these are land registry documents, building notice certificates and architectural drawings of the house in question. But no – I soon realise that all these refer to the very house we are sitting in and not to the one we want to buy. What is going on here? Furthermore, when I am brought into the conversation, I am told that they have indeed been talking about the weather and how brave Uncle is not to mind riding around viewing houses in the driving rain. Prakash's wife and Anish then have a little private conversation, which also seems to be something to do with me. Anish then starts poking my stomach – apparently she was expressing concern about how thin I am. Quickly gathering up the papers, she stuffs them back into the wardrobe and hurries off to the kitchen to provide tea and biscuits. So in terms of a negotiation we have not really got started.

It transpires that we will have to go and discuss the matter with the agent face to face. So next day, again in pouring rain, we set off on our bikes to do battle with him. His office is down a muddy back street and we only know which one it is because Prakash is outside waving to us. We enter the building, which is nothing more than a dingy hut and are shown into a bare room, where a single light bulb hangs from the ceiling. Old wooden benches run round the wall for us to sit on and there is a plain desk, behind which sits 'Mr Big'. He is wearing Army Combos, a peaked khaki cap and is tapping fiercely on a pocket calculator, occasionally making pencil notes on a pad of paper. The dingy atmosphere of the room makes me feel we are entering into a firearms deal in some rebel-torn republic, or involved in trying to overthrow some unscrupulous dictator. I half expect to see armed guards with Kalashnikov rifles at the ready.

Mr Big looks up as we enter, slams down his calculator and greets us in turn with a meaty, bone-crunching handshake.

"Sit down, sit down. Garcia, get some tea for the boys" – or words to that effect, and for the next hour I understand not a word as he hardly draws breath, occasionally allowing our ringleader Ramesh to interject briefly before continuing with his barrage of well-rehearsed, aggressive counter-arguments. (Meanwhile my imagination plays on: I hear a distant explosion, then some rapid gunfire followed by a rebel soldier staggering into the room, clutching his side in agony before collapsing in a pool of blood on the floor.)

In fact, by Nepalese standards this is probably quite low-key bargaining and we end up by getting a bit off the price, shaking hands and having a friendly chat over a cup of tea. Prakash will bring all the paperwork to Wasta Care Centre early tomorrow morning and we will formalise the deal. We are shown out into the wet muddy side street; the rebel forces have fled and peace has returned to Kathmandu.

In the afternoon, Ashish, Ramesh and I set off to spend the rest of the day and night at Rajesh's house in Budhanilkantha. It is raining harder than ever, so the plans to do some painting in his garden have been shelved. Instead, we (including Soni and Basanta) set off for brother Rhadeshyam's new house near Rajesh to spend the wet afternoon in his recording studio, playing music and singing songs. I have brought with me the sheet music for Paul McCartney's song "Yesterday", hoping that it might appeal to them. The Mersey sound is a far cry from the Nepali sound and they only have a dim awareness of who the Beatles were, but after one or two tries I can tell they are beginning to warm to the tune and the lyrics. We do several runs-through and before too long are ready for a take. I have already made a recording of the piano accompaniment and Ashish has added a string backing.

The five of us make the recording, listening to the accompaniment through earphones. By now, they have taken the song to their hearts and are clearly enjoying it. This is possibly the first time that "Yesterday" has been sung by a Nepalese choir and the result is quite beautiful. I am proud of them and Ashish promises to post it on YouTube.

By the time we return home, Sunil has managed to do some research into recent property sales. Being a resourceful young man with a wide circle of friends, he has learnt that we are being hugely overcharged on house number four.

"When you come looking like a wealthy Westerner, Uncle, immediately all the prices go up."

So I promise to take no further part in any proceedings. Now that the boys have some idea of what they are looking for and know how much I am prepared to pay, I suggest that they spend the rest of their holiday after I have gone home looking around and testing the market.

"There seems to be a lot of property available so don't rush into anything," I say. "And keep me informed."

They agree to do this when we return from our trip to Pokhara, which is to take up my final three days.

# Chapter Twenty-Three

Each time I go to Kathmandu I find myself becoming better acquainted with one or two other members of the family. This time it is Karjun's sons, Ashish and Abhishek. I have mentioned earlier how close Ashish is to the orphans. In fact he spends more time with them than he does at his own home. He eats and sleeps there and so is totally involved in their lives. He is twenty years old, and in contrast to them, is totally uninhibited, extrovert and exuberant. He jokes, he sings, he plays the guitar, writes songs, directs films with his friends and above all has a great sense of humour. He loves his course at Catering College and is looking forward to six months in Thailand working in a five-star hotel. He comes out with statements such as "It's not a matter of what can I do in life, Uncle. It's more a question of what will there be left for me to do in life." So he is like a breath of fresh air whenever he comes into the room, because he quickly turns an atmosphere which is gloomy or serious into one which sparkles with life. If he has a fault, it is that he spends too much of his time in front of the mirror, preening and admiring himself. I tell him the tragic story of Narcissus, the boy in Greek mythology, who drowned himself admiring his own reflection in the river– but he shrugs it off: "I know how to take good care of myself."

Ashish even went so far as to write a character description of himself under the title "Beyond", which is his pseudonym. These are his words:

*Beyond – just a character developed by a funny and unexpected situation of a moment when I was making my hair in the mirror. The print on my t-shirt said "Beyond the shadow of doubt". Then the word "Beyond" suddenly got my attention. And I'm like, "Hey, wait a sec, the meaning and my way of living is the same." From that moment a new artistic name was tagged before my personality.*

*Well, I am just a non-serious guy or we can say I just love to smile and live a life which ain't about sadness and fear, it's about smile and confidence.*

*I've faced many situations in life. Many moments that drowned me in tears, many moments that crushed me, tore my heart apart. But then I remember, no matter what the problem is, it won't bypass you. Actually, even if you are sad or unhappy about it, it will come to you no matter what. So why not face it with a smile? That is me.*

*Music, dancing, singing – well, others may take it as some entertaining factor or some other stuff, but I believe in them. I can be a new me, a creative me. For they are my passions. Actually, whatever amazes my eyes becomes my passion and I start working on that."*

Abhishek is only fourteen but shares his brother's impish sense of humour and creativity. He clicked endlessly on the trip to Pokhara and with the various functions on his mobile, would add colours and special effects to his pictures giving them a special individual touch. Whenever he makes a joke, usually at my expense, he collapses in paroxysms of laughter. I got to know Abhi really well on this trip because he would cling to me like a puppy, constantly asking questions, teasing me and always clicking me at quite unclickable moments, such as when I was sleeping or eating – he thought this was hilarious. He gave me three of his comic books when I left. He obviously thought I needed to laugh more!

Nothing is ever plain and ordinary when these two brothers are around and they make endlessly cheerful company on our trip away.

Pokhara is the trekking capital of Nepal and, as such, is the country's most popular tourist destination after Kathmandu. The bus trip takes between six and eight hours, depending on the current state of the roads, but one can fly. A small airport was built in the mid 1950's, which considerably increased the number of visitors in the late 1960's and early 1970's when it became particularly popular with the hippie community. The Pokhara valley is situated in the centre of Nepal and is hemmed in closely by the giant peaks of the Himalayas. I had heard about the town's beautiful lakeside setting and its fame as the starting point for treks through the mountains of the Annapurna Range, so it was a long-held ambition of mine to visit it with some of the family.

Several weeks before my arrival Rajesh set about making arrangements. October is a good time to visit, being the start of the dry season. With the monsoon only recently finished, the climate is still warm, the countryside green and lush and the conditions just right for walking and climbing.

We are a party of thirteen that assembles at Kathmandu Airport. Ira is not able to come so her daughter Indu joins the party instead. Most of them have never flown before, so there is a terrific sense of excitement and nervousness. It is a thirty minute flight and we are to go by Buddha airlines. Clicking starts to take place from the moment we gather in the departure lounge and continues as we walk across the tarmac, climb the steps into the plane and especially once we are airborne and above the clouds, enjoying the magnificent views of the Himalayas. It is a thrilling experience for us all, and the flight is over all too quickly. On arrival at Pokhara airport, we assemble just outside the terminal building for a group photo before a minibus takes us into the town.

We are staying at the Hotel SplendidView and it certainly lives up to its name with its balconies giving great views of the mountains and Lake Fewa. Rajesh has arranged for all our meals to be taken at the restaurant on the other side of the road and Kaji the guide has planned our activities. After lunch we spend the rest of the first day walking around the town, watching boating activities on the lake, and visiting two local sights: Devi Falls and Siva Gupa Cave and Grotto. Pokhara is the most popular town in Nepal and one can understand why. Its lakeside setting is beautiful, while the mountain ranges of the Annapurna are spectacular against a clear blue sky.

We are not here for long enough to do any serious trekking and anyway it is much too hot. So the next morning we hire bicycles and have races along the northern shore of Lake Fewa. The bikes are rickety and the roads are poor, but everyone has a great time. For most of the party, this is the first real holiday of their lives – a completely new and exciting experience. We return to our restaurant for a well-earned lunch. Again, to be served such a delicious variety of food in endless quantities is something quite new for them. We all retire to our rooms for a long siesta in the heat of the day.

Kaji has arranged for us to go boating that afternoon, so we meet at the lakeside at about 4 o'clock. We take out three boats with a steersman, but we are all allowed to take turns with the paddle. We launch into the lake, steering towards the little island from where we can hear the tinkling prayer bells of a small Hindu Temple. There is some hilarity – and of course a lot of clicking – as we each try our hand with the paddle, making the boat sway uneasily from side to side as we change places.

"Ours is called 'The Unsinkable,'" I announce across the water. "What is yours?"

There is a moment's thought, then Ramesh calls out, "Titanic!"

"And yours?"

"Submarine!" announces Niran.

We have paid for an hour's boating but Mama is enjoying it so much that she insists on another hour. Some people sing boating songs; others strike poses for camera shots against the mountain backdrop. Gradually, the clouds above us become tinged with pink and bright shafts of sunlight shine through gaps between the grey clouds to the west. We head back towards the jetty as the last of the daylight fades from the sky.

In the evening everyone is happy either to watch TV or play games (I teach them a card game called 'Fish' which they play riotously and very competitively), and there is a chess set which also provokes fierce competition. They have different names for some of the pieces in Nepal. For instance, the Queen is the Prime Minister, the Bishop is the Camel, the Castle is the Elephant (which suggests that 'castling' is 'elephanting' – but the boys don't know that move so I have to teach it to them) and Pawns are Policemen. It is all very quaint and charming. They tend to gang up in pairs against me, discussing tactics loudly, knowing that I can't understand them. Their joy is unsurpassable if they beat me – and whenever I get close to beating them, they cheat shamelessly which they also find hilarious. It is very provoking.

I ask Indu if she is happy being the only girl among all these boys.

"Very happy, Uncle. They are nice boys and really are my brothers in our family, so we have all grown up together. They are good friends with me."

Indu is 17 and will soon be starting at college to do a nursing course. She is a confident girl, easily able to hold her own in this predominantly male company.

For our final day, Kaji has arranged that we should see the sunrise from the top of Sarangkot Mountain. This will mean an early morning call at 4:00 am, so we turn off the TV, end our games early and are in bed by about 10:00 pm.

I sleep very well and in fact am awake before the early morning call. I get up and go around knocking on doors to try to raise the sleepy heads. Gradually they haul themselves out of bed, bleary-eyed, throw on some clothes and stagger into the minibus waiting below. There is silence apart from a few cocks crowing, and it is pitch black. The general mood is sullen. Nobody speaks and our spirits are not lifted by the rainy weather and the bumpy uphill drive. We pass the occasional jogger and soon encounter other motorists eager to make the summit by sunrise. The sight of these travellers at this time of day is encouraging. Maybe it is going to be worth the effort after all.

It is still pitch black when we park and clamber out of the minibus. All we can see are the twinkling lights of Pokhara in the valley far below. Kaji leads us up a few short flights of steps to the viewing point and there we join other sunrise-seekers, expectantly waiting, cameras in hand. There is a gap in the clouds above the horizon and a pale light from the east begins to filter through. We are starting to feel hopeful that something spectacular will soon happen in the sky above.

Within half an hour, our viewing point and the nearby hotel look-out station are thronged with hopeful visitors. Gradually, the grey clouds become tinged with pink, soon turning to a red stain which grows and expands upwards above our heads. Banks of white clouds become visible in the valleys below, and behind us tall shafts of misty rain are twisted by the breeze as they fall to earth.

Now the day is beginning to break and dark distant shapes turn into green fields and trees. But where exactly is the sun? Are the clouds going to cheat us of our hopes, or will they break apart so that the sun can make a magnificent entrance?

"Look, a small red glow above the horizon, Uncle!" Arun whispers to me. I look in the direction he points and train my eyes on the spot of colour.

As the red brightens into orange a few cheers are heard and moments later we see the dazzling edge of the sun's disc, rising from behind the farthest mountain. We all feel a tremor of excitement. The ever-increasing glow burns yellow upon us, giving extra lustre to our faces. As Ashish turns his camera to take a picture of himself, he spots something in his viewfinder.

"A rainbow! Look behind."

We all turn and see a perfectly coloured arc towering over our heads and the distant hill. In fact it's a double rainbow. How auspicious is that! We hardly know which way to turn: on one side the ever brightening rising sun and on the other, the magnificent rainbows glowing multicoloured against the grey rainclouds. We simply bask in the colourful show, enjoying its effects on land and sky.

I had long dreamt of ending the account of these experiences with a moment like this, sharing some special wonder of Nepal with my new-found family. So my dream came true on the summit of Sarangkot Mountain at the break of day, high above the lakeside setting of Pokhara. It was a magical ending to our holiday.

To find out more about Nick, the young people he supports in Nepal and how you can help please visit **www.discoveredinkathmandu.co.uk**

# Afterword

The boys now have their house. It has nine rooms on two and a half floors, and when they have fitted the ground floor with a kitchen, they will let it out (I gather it is already booked). It is in an area called Imadol, only twenty minutes from Pulchowk so Ashish and Abhishek can be frequent visitors and occasional residents.

I understand that Kiran's tailoring business is expanding well, Santosh has started studying in Bangladesh and Indu has begun her nursing course at Vinayak College of Science and Health.

Their lives are a constant source of interest to me, and I always look forward to my annual visit with great anticipation. I really cannot decide who is the more fortunate, them or me, to have made this discovery in Kathmandu.